Fire Safety in Buildings Questions and Answers

Other titles in the series:

CDM 2015 Questions and Answers: A practical approach to design, safety and wellbeing, Fourth edition (2021)
Pat Perry. ISBN 978-0-7277-6584-0

Health and Safety: Questions and Answers: A practical approach, Second edition (2016)
Pat Perry. ISBN 978-0-7277-6074-6

Risk Assessments: Questions and Answers: A practical approach, Second edition (2016)
Pat Perry. ISBN 978-0-7277-6076-0

emerald PUBLISHING

ice Publishing

Fire Safety in Buildings Questions and Answers

Pat Perry

CMIOSH

Published by Emerald Publishing Limited, Floor 5, Northspring, 21-23 Wellington Street, Leeds LS1 4DL.

ICE Publishing is an imprint of Emerald Publishing Limited

Other ICE Publishing titles:
Designing a Safer Built Environment: A complete guide to the management of design risk
John Carpenter. ISBN 978-0-7277-6582-6
Refurbishment Projects: Health and safety management
Edited by David Oloke. ISBN 978-0-7277-6159-0
Practical Guide to Using the CDM Regulations 2015: Teamwork not Paperwork
Tony Putsman and Paul McArthur. ISBN 978-0-7277-5990-0

A catalogue record for this book is available from the British Library

ISBN 978-0-7277-6647-2

Cover photo: From left to right: (1) The Address Downtown hotel in Dubai. arabianEye FZ LLC/Alamy Stock Photo. (2) An engineer checking an industrial fire control system. A_stockphoto/Shutterstock. (3) Firefighters spraying water to extinguish a fire in a burning building. Ponpun Wichainsan/Shutterstock

Commissioning Editor: Michael Fenton
Assistant Editor: Cathy Sellars
Production Editor: Sirli Manitski

Typeset by Manila Typesetting Company
Index created by LNS Indexing
Printed and bound by CPI Group (UK) Ltd, Croydon, CR0 4YY

Contents

Preface

Fire Safety in Buildings: Questions and Answers has been written as a practical guide to the wide-ranging subject of fire safety.

This book follows the very successful format of practical questions and answers found in *CDM 2015 Questions and Answers: A practical approach to design, safety and wellbeing* (2021), *Health and Safety: Questions and Answers: A Practical Approach* (2016) and *Risk Assessments: Questions and Answers: A Practical Approach* (2017).

The book is aimed at a wide readership from clients, employers, developers, architects, construction professionals, property managers, students and anyone who has responsibility for buildings where people are employed, live or visit. It is a quick 'pocket book' guide to the key provisions of fire safety law and is intended to give an overview of fire safety and to direct the reader to more detailed information where appropriate.

It is a practical guide to the subject of fire safety. It is not a detailed technical handbook of fire safety engineering, fire safety law nor fire safety science.

Fire safety has always been an important topic. Sadly, changes to the law and regulatory standards only occur when major incidents happen. There have been some catastrophic incidents in recent years and the law of fire safety has changed.

Prior to the Regulatory Reform (Fire Safety) Order 2005 (RRO), fire safety law was spread over 80 acts of parliament and regulations. The RRO consolidated the laws and is still the basis of substantial fire safety law today.

The Grenfell Tower fire in 2017 heralded new fire safety and building laws governing high-rise residential buildings. In addition, amendments have been made to the RRO and new regulations have been introduced to strengthen fire safety.

This book covers those new requirements for high-rise buildings and explores the new duties placed on responsible persons and others involved in the design, construction and management of buildings.

Each chapter of this book is designed to stand alone and to be used as a reference guide to provide the reader with quick access to pragmatic information on how to comply with fire safety law and to understand the key principles of fire safety and prevention. Given the interconnections between the subject areas, some topics are mentioned in more than one chapter.

Fire safety is about prevention – preventing fire from starting, fires from spreading and fire preventing escaping from buildings and other structures.

Often, fires become catastrophic because of a failure in the first line of defence. There is then a knock-on effect as other precautionary standards fail. Risk is interconnected. An error, oversight or event in one area may trigger an event in another area. Grenfell Tower took the lives of over 70 people because, amongst other reasons, a fire started, it was not contained, it spread to flammable materials, it breached compartmentation, it took advantage of poor construction methods, escape routes were inadequate and residents failed to know what to do to protect their own and others' safety.

Failures by building owners, building managers, designers, contractors, residents and the fire service.

Knowledge is important. It is hoped that the contents of this book will add to the reader's knowledge of fire safety and provide them with a thirst to seek additional knowledge on the subject of fire safety.

Pat Perry CMIOSH
May 2023

About the author

Pat Perry CMIOSH qualified as an environmental health officer in 1978 and spent the early years of her career in local government in the UK enforcing environmental health regulations, particularly health and safety law, which became her passion.

Pat left local government and established a very successful Environmental Health Consultancy which was subsequently integrated into a large regulatory compliance group.

She has extensive knowledge of the retail, commercial and hospitality sectors and has served on various working parties on both health and safety and food safety issues. Pat regularly contributes to professional journals and appears as a speaker at related conferences and seminars. Her previous publications include *CDM 2015 Questions and Answers*, *Health and Safety: Questions and Answers*, *Risk Assessments: Questions and Answers* and *Fire Safety: A Practical Approach: Questions and Answers.*

Perry P
ISBN 978-0-7277-6647-2
https://doi.org/10.1680/fsbqa.66472.001

Chapter 1
Introduction to the principles of fire

What is fire?

A standard definition of fire is

> a process of combustion characterised by the emission of heat accompanied by smoke or flame.

Mostly, when the term fire is referred to, we usually mean the uncontrolled development of fire not a controlled fire such as happens in furnaces and so on. Fire is often unwanted, unexpected, disastrous and costly, both in terms of human life and business costs.

What causes a fire?

There is an inter-relationship between heat, fuel and oxygen. This is known as the 'Fire Triangle'.

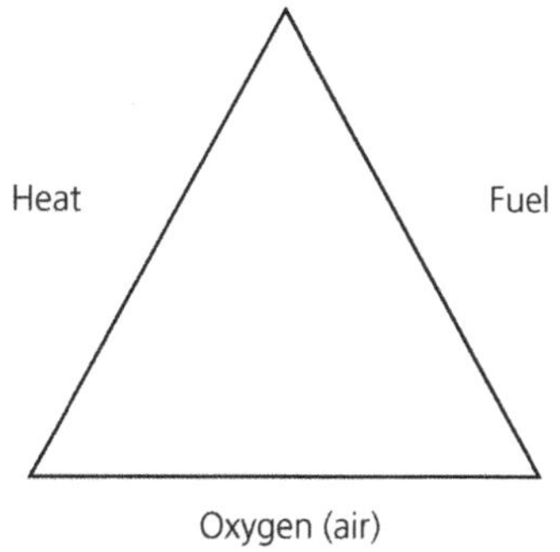

Fire is the result of a chemical reaction between a fuel and oxygen. Fire cannot occur if one of the key components is missing – that is, eliminate either the heat, the fuel or oxygen.

This is the principle under which fires are fought

- drench with water thereby removing the heat
- smother with inert material – for example, sand to deprive the fire of oxygen
- remove the fuel – for example, separate burning material from other materials.

What are the four stages of fire?

It is generally accepted that the four stages of fire are

- ignition/incipient
- growth
- fully developed
- decay.

Ignition/incipient

This first stage begins when heat, oxygen and a fuel source combine and have a chemical reaction resulting in fire. This is usually represented by a very small fire, which often goes out on its own before the following stages are reached. Recognising a fire in this stage provides your best chance at suppression or escape.

Growth

The growth stage is where the structure's fire load and oxygen are used as fuel for the fire. There are numerous factors affecting the growth stage, including where the fire started, what combustibles are near it, ceiling height and the potential for 'thermal layering'. It is during this shortest of the four stages when a deadly 'flashover' can occur, potentially trapping, injuring or killing building users, residents and firefighters.

Fully developed

When the growth stage has reached its maximum and all combustible materials have been ignited, a fire is considered fully developed. This is the hottest phase of a fire and the most dangerous for anybody trapped within the building.

Decay

Usually the longest stage of a fire, the decay stage is characterised with a significant decrease in oxygen or fuel putting an end to the fire. Two common dangers exist during this stage

- existence of non-flaming combustibles can potentially start a new fire if not fully extinguished
- danger of a backdraught when oxygen is reintroduced to a volatile, confined space.

Is it the actual material that burns?

In many cases it is the body of the material that burns, for example, furnishings, wood, clothing and construction materials, but in many instances it is also the gaseous vapours that are given off during the burning process re-igniting and causing the fire to develop.

How does fire spread?

Fire generally spreads in six ways.

Direct contact

Materials adjacent to each other ignite through a domino-type effect.

Radiation

When a fire is strong enough, it can emit sufficient radiated heat to ignite combustible materials that are not in direct contact with it. This process is similar to sunlight directed through a lens onto dry paper or straw – the paper or straw will eventually ignite.

Conduction

Heat is transferred through materials that ignite adjacent materials that are in contact. A wall may get so hot that it ignites the surface coverings on the other side. Handles on fire doors often get so hot that they cannot be touched, thereby confirming a fire inside the room with very high temperatures.

Convection

Currents of air rise above a fire and may concentrate at an upper limit – for example, beneath a ceiling. When substantial hot gas and smoke has accumulated, the temperature at the upper level rises to such an extent that it can set materials on fire, thereby causing a secondary fire and possibly a flashover (see below).

Flashover

The hot air and gases accumulated by convection may eventually start to descend to a lower level through radiation and, at sufficiently high temperatures (around 500°C), may make combustible materials start emitting gases that spontaneously ignite.

Backdraught

Fire in a room with little oxygen will start to decay but, if a door opens or glass in a window shatters, the sudden inflow of air/oxygen can reignite the fire with explosive effect.

How quickly does fire spread?

Fire development is often described as exponential. In simple terms, fire tends to double its size at regular intervals (sometimes as quickly as every 30 s) causing everything in a room to burn in as little as three minutes.

Fires in buildings develop differently to fires outdoors – for example, bonfires and the speed at which they spread takes most people by surprise.

The rate of fire spread within a building will depend on

- the degree of fire separation (fire compartmentation) that exists
- the combustibility of the materials present (some materials such as soft furnishings may burn at faster rates than others)
- the presence of air/oxygen (draughts and convection currents can fuel fires to burn more rapidly)
- the presence of active firefighting systems to detect and suppress fires.

It is said that the average domestic sitting room with usual equipment of TV/video and furnishings will be totally alight in about 4 min.

Often, members of the public fail to comprehend the speed at which fires develop and do not believe that they must evacuate or do anything in relation to their safety.

The major fires in Woolworths, Manchester in 1979, the Dublin Disco in 1981 and the Bradford fire in 1985 are all examples of a rapidly spreading fire and people delaying in evacuating the building.

What is a wind-driven fire?

A wind-driven fire may be described as one where external wind or ventilation-forced pressure causes strong air movements, affecting the severity of fire spread. It is often used to describe fires that may also be referred to as forced-draft, wind-assisted, force-vented or blowtorched.

Wind-driven fires can occur at ground floor level. High-rise buildings, by their very nature, create abnormal, unpredictable air flows and wind speed generally increases with height.

Fires can be affected by wind pressure and high-velocity air movements. The impact can be experienced in open fires or wildfires, while in buildings the greatest impact is usually experienced with fires in tall buildings. If windows are open, or have failed through exposure to heat, external wind can affect the speed and direction of fire development.

In all instances where there is an oversupply of oxygen then

- the fuel will burn at a higher temperature
- the fuel will burn more quickly
- the fuel liberates energy at a higher rate.

By definition, a higher energy fire will be harder to put out, usually requiring more water and some form of direct attack.

The Grenfell Tower fire was fatal for so many people because of external spread of flame. What does this mean?

External fire spread may develop if the fire breaches the external envelope of the building. This may be a result of external window glazing or wall panels failing, or by way of open windows. External fire spread may compromise compartments above or below the fire floor, irrespective of the materials used in construction.

A fire can spread from combustible material to combustible material or be spread by combustible gases. Flames may take hold in external cladding, as in Grenfell Tower, and travel rapidly across the material. Cavities may exist as part of a cladding system or facade assembly. Cavities may also be created by damage, delamination or movement. Fire may spread within cavities, with flames becoming elongated due to the restricted space. The flame length may increase regardless of the materials used in construction, resulting in rapid, hidden fire spread if appropriate fire barriers are not present.

Fire travels up the building in between the external cladding, travelling along the cavities and transferring to new combustible material.

Why is smoke such a hazard?

Smoke is defined as

> the product of combustion, consisting of fine particles of carbon and other substances carried by hot gases and air.

Smoke spreads easily around a building, and scientifically it is proved that smoke moves faster than people can walk.

Smoke contains particulate matter, and this makes the 'flame' opaque – so you can't see through it. Therefore, it is a hazard because it cuts down visibility.

Smoke also contains the 'products of combustion', depending on what materials are burning, and contains poisonous gases and toxic fumes. Common gases produced during combustion include carbon monoxide and hydrogen cyanide.

Carbon monoxide, an odourless colourless gas, is lethal and kills within minutes at high concentrations. Most people caught in a fire die from breathing in carbon monoxide gas.

Before inhalation of the smoke becomes lethal it can cause severe respiratory irritation and breathing becomes difficult.

Carbon monoxide, in non-lethal concentrations, will affect a person's ability to concentrate. Combine a lack of concentration with visual difficulty, and quickly people become disorientated in a smoke-filled environment. This is one of the reasons why fire exit and emergency signage must be clear and properly lit.

People are reluctant to walk through smoke even though it would lead them to safety. People therefore become trapped, perhaps by their own making, and as a result suffer injury or death.

What does the term 'flashover' mean?

Flashover is defined as

> a sudden transition to a state of total surface involvement in a fire of combustible materials within a compartment.

After the first item in an enclosure is ignited, hot gases rise vertically in a narrow plume into which air is pulled from the natural ventilation in the room, so increasing the volume of the smoke and gases.

As smoke reaches the ceiling it spreads out in all directions and begins to form a thickening layer below the ceiling.

As the fire grows, the flames reach the ceiling and are deflected horizontally across the surface of the ceiling. Radiation is so strong that flame will spread over all combustible surfaces and items over a relatively large area, reaching temperatures at which they spontaneously burst into flames.

At flashover, virtually all items in the room are alight because the room temperature exceeds 1000°C and survival of occupants in the room is impossible.

What does the term 'backdraught' mean?

Backdraught is defined as

> a phenomenon in which a fire that has consumed all available oxygen suddenly explodes when more oxygen is made available, typically because a door or window has been opened.

A backdraught is where limited ventilation can lead to a fire in a compartment producing fire gases containing significant proportions of partial combustion products and unburnt pyrolysis products. Pyrolysis is the chemical decomposition of organic (carbon-based) materials through the application of heat. If these products accumulate, the admission of air when an opening is made to the compartment can lead to a sudden deflagration (to burn rapidly with intense heat and sparks being given off). This deflagration moving through the compartment and out of the opening is a backdraught. The force of a backdraught has the potential to damage building elements, resulting in an unstable structure.

Case study: King's Cross fire (1988): Flashover

The fire brigade was called at 19:36 to a fire on an escalator at King's Cross Underground station. The fire seemed controllable and in one location. Passengers were evacuated up escalators by way of the ticketing hall. Tubes were still stopping and passengers were still embarking and disembarking.

At 19:45 the fire suddenly erupted up into the ticket hall. It was described as 'flashing, with searing heat and thick smoke through the ticket hall'. And 'within seconds, the area was in total darkness and the conditions had become unbearable'. Escalators carried passengers up into the ticket hall and certain death.

The cause of the fire was established to be discarded smoking material that fell through a wooden escalator onto grease and debris that was very combustible.

As a result of the King's Cross fire, 31 people died and many more suffered catastrophic burns and other injuries.

Extract from the Public Inquiry Report into the King's Cross Underground fire (published in November 1988)

20. From 19:43 there was a rapid worsening of the conditions in the tube lines ticket hall. At about 19:45 there was a sudden eruption of black smoke and flames into the ticket hall. The flashover had taken place. The time of 19:45 shown on the stopped digital clock was confirmed by P.C. Dixon who, having escaped via the exit to the south side of Euston Road, and after leading P.C. Hanson to the street, immediately radioed his headquarters for assistance and also informed them of a 'major incident' at King's Cross Underground station. This message was recorded at 19:46:03 (i.e. 19:45:58 after being corrected] which, allowing time for P.C. Hanson to escape, suggests the time of flashover at about 19:45. So in a period of about two minutes or less, the fire observed by the firemen on their first arrival had deteriorated from what they perceived as a modest fire into a raging inferno.

21. Very few people who were in the tube lines ticket hall and who witnessed the flashover survived, and most of those who did survive were seriously injured. Others coming up the Victoria Line escalators had a limited view of the flashover looking up the escalator shaft. Temporary Sub-officer Bell who was at the bottom on the Piccadilly Line escalator shaft probably saw the fire a minute before the flashover and at some time shortly after the flashover.

22. Shortly before the flashover P.C. Hanson was a short distance down escalator 9 on the Victoria Line urging people up the escalator and staircase. He became aware of dense smoke in the ticket hall so he went to investigate. When he got about five feet into the ticket hall he saw '. . . what I can only describe as a large wall of flame or fire. It was definitely above head high, and immediately following this was like a whoosh . . . and a large ball of flame, which was about head height, hit the ceiling in the ticket hall itself. This was followed almost instantaneously by dense black smoke . . .'

P.C. Hanson later amplified this:

'To be more accurate I would say it was a jet of flame that shot up and then collected into a kind of ball' and then 'I saw it shoot up across the top of number four and collect along the roofing . . .'

What is flame spread?

Flame spread is continuous ignition. Flames spread over a solid surface as the surface in advance of the flame becomes heated by radiation, conduction or convected heat transfer, causing gaseous fuel to be released and so extending the burning zone.

The speed of advance offlames can be increased by air movement – for example, a flame driven sideways by the wind extends the area of heating, thus causing more heat, gases and fumes.

Flame spread is much quicker when the surfaces' temperatures have been pre-heated and rapid flame spread can lead to flashover.

What is spontaneous combustion?

Encyclopaedia Britannica defines spontaneous combustion

> as the outbreak of fire without application of heat from an external source.

This combustion can occur when flammable matter like oily rags, damp hay, leaves or coal is stored in bulk. Spontaneous combustion (sometimes called spontaneous ignition) begins when a combustible object is heated to its ignition temperature by a slow oxidation process. Oxidation is a chemical reaction involving the oxygen in the air around us gradually raising the inside temperature of something – for example, a pile of oily rags – to a point at which a fire starts.

The way combustible materials are stored has a lot to do with whether they will spontaneously combust or not. Aerosols and other flammable liquids or gases often have instructions about being stored away from a heat source. This prevents temperatures rising to the point of ignition.

Substances with low ignition temperatures pose a greater risk than others.

Flames, heat and smoke are major hazards in a fire. What else causes a potential risk?

When materials burn, they create 'products of combustion'. These will include various gases that could prove fatal if inhaled.

Carbon monoxide gas can be produced during a fire. Carbon monoxide is hard to detect because it has no smell, taste or colour. This also means that it is easy to inhale without realising.

Carbon monoxide is produced when fuels such as gas, oil, coal and wood do not burn fully.

When a fire burns in an enclosed room, the oxygen in the room is gradually used up and replaced with carbon dioxide. Following a build-up of carbon dioxide in the air, the fuel is prevented from burning fully and it starts to release carbon monoxide.

Hydrogen cyanide (HCN) is produced when materials that contain nitrogen in their structure – for example, orlon, nylon, wool, polyurethane, urea-formaldehyde and ABS (acrylonitrile-butadiene-styrene) – are involved in the fire. HCN and other cyanogen compounds arrest the activity of all forms of living matter. They exert an inhibiting action on the use of oxygen by the living cells of the body tissues.

Hydrogen chloride (HCl) is produced when polyvinyl chloride is decomposed by fire. If inhaled, HCl will damage the upper respiratory tract and lead to asphyxiation or death.

There are three common oxides of nitrogen: nitrous oxide, nitric oxide and nitrogen dioxide, of which there are two forms. Nitrogen dioxide, which is very toxic, can be produced from the combustion of cellulose nitrate. Nitric oxide does not exist in atmospheric air because it is converted into dioxide in the presence of oxygen. These compounds are strong irritants, particularly to mucous membranes and thus, when inhaled, will damage tissues in the respiratory tract by reacting with moisture to produce nitrous and nitric acids.

Sulfur dioxide is an inorganic irritant gas often produced on the combustion of fossil fuels, but it may also be produced on the decomposition of any sulfur-containing compounds, such as vulcanised rubber. Sulfur dioxide is a respiratory irritant and may restrict people's breathing, especially those with lung conditions such as asthma.

Other substances that could be produced during combustion include ammonia, chlorine gas, formaldehyde, phosgene and many inorganic compounds.

Toxic gases and vapours produced by combustion are responsible for the majority of deaths at building fires.

Why do buildings often collapse in a fire?

At some stage in the development of fire, the walls, floor and ceilings of a building are subject to heating, mostly in the form of radiation, at temperatures of many hundreds, if not thousands, of degrees.

The capacity of a building to withstand this heat and still support loadings and prevent the passage of heat and gas diminishes with the period of exposure to sustained heating.

Steel girders can buckle in the heat and lose their load-bearing capacity. Other materials burn to ashes and fail to support loads.

The structural integrity of a building during a fire is a critical part of fire protection and without such fire protection, buildings would be extremely vulnerable to collapse.

What is meant by the classification of fires?

Fires are classified according to ISO 3941 (2007), a harmonised European standard and are classified so that the suitability of different extinguishing agents can be determined.

The classification depends on the actual type of combustible material involved in the fire.

Class A	These fires involve solid material, usually of an organic nature, in which combustion normally takes place with the formation of embers. Material types are usually wood, fibre, textile fabrics and paper. Cooling by water is the most effective method of extinguishing this type of fire.
Class B	These fires involve liquids or liquefied solids such as petrol, oil, greases and fats. The most effective way of extinguishing this group of fires is by smothering the fire and depriving it of oxygen. Types of extinguishing agent would be foam and powder. A fire blanket can also be used in certain situations.

Class C	These fires involve gases such as hydrogen, propane, butane and other liquefied petroleum gases. Gaseous fires are extremely difficult to contain, and the most effective way is to cut off the gas supply at source. Any burning material set alight by the fire can be dealt with by the best medium.
Class D	These fires involve metals and other substances such as magnesium, sodium, aluminium, potassium, calcium. Special techniques for firefighting are required.
Class F	Class F fires consist of cooking oils and fats that have been ignited. The high flash point of cooking oils and fats and the extremely high temperatures necessary to cause a blaze with these materials has led to their a separate classification. These kinds of fires are most common in restaurants and commercial kitchens. A common cause of Class F fires relates to deep fat frying; oil can be left to cook for too long or in too great a quantity. Deep fat frying is the leading cause of accidental fires in kitchens, so great care is always required. Most often, they occur when pans containing oils are left unattended or are not carefully monitored. Signs of a potential Class F fire, such as seeing or smelling smoke, means you should turn the heat off immediately before the oil reaches too high a temperature and ignites. Prevention is incredibly important, and busy environments like large commercial kitchens should enforce strict fire safety standards. A fire blanket is often the best solution for smaller cooking oil/fat fires, eradicating the need to clean up any debris from the use of chemical suppressants. The appropriate extinguisher used to tackle a Class F fire is a wet chemical extinguisher.

What is an electrical fire?

Strictly speaking there is no such thing as an electrical fire.

Electrical fires are not given their own full fire class, as they can fall into any of the other classifications. After all, it is not the electricity that is burning but the surrounding material that has been set alight by the electric current.

An electrical fire is often used to describe a fire in which electricity is present. The electricity may be supplying the source of heat (ignition) through electrical sparks or overheating, and the fuel may be the electrical cabling and wiring.

The fire in an electrical appliance will fall into whatever category the burning material is.

But water should never be sprayed on to live electrics, so water-based fire extinguishers are not suitable on live electrical fire but could be used once the electrical supply has been switched off.

However, the most suitable fire extinguishing medium to use on live electrical equipment is carbon dioxide or a powder fire extinguisher.

What are some of the major causes of fire?

Some of the most common causes of fire are

- faulty electrical equipment
- smoking and smoking materials
- fat fryers and chip pans
- portable heaters
- refuse/rubbish accumulations
- flammable materials.

Arson is now a common cause of commercial fire and incidents are rising in respect of domestic premises. Easy access to ignition sources and fuel make arson easy. The number of school fires due to arson is increasing. Preventing access to grounds and buildings and avoiding accumulations of combustible materials will help prevent arson.

What are the statistics regarding fires in the UK?

Annual fire statistics were published for England by Gov UK, in November 2022, for fires and related incidents for the period from June 2021 to June 2022.

In that twelve-month period, local fire and rescue services attended a total of 154 772 fires.

Of the 154 772 fires, there were 63 751 primary fires – that is, those occurring in non-derelict buildings, vehicles or outdoor structures, and involved a fatality, casualty or rescue or involved more than five pumping appliances.

There were 751 fires in purpose-built high-rise (ten or more storeys) flats/maisonettes.

There were 275 fire-related fatalities, of which 208 related to dwelling houses.

There were 6192 non-fatal casualties, of which 2526 required hospital treatment.

Fire and rescue services attended 233 471 false alarms during the year.

KEY RESOURCES

ISO 3941:2007 Classification of fires

Public Inquiry Report into the King's Cross Underground Fire (1988)

Department of Transport. Fire and rescue incident statistics: England, year ending June 2022, Published 10 November 2022

Perry P
ISBN 978-0-7277-6647-2
https://doi.org/10.1680/fsbqa.66472.013

Chapter 2
Fire safety legislation

What is the main piece of legislation governing fire safety in buildings?

The Regulatory Reform (Fire Safety) Order 2005 (RRO) contains the legal requirements for managing fire safety in premises and is the main legislation applicable to most buildings.

Further to the Grenfell Tower fire tragedy, numerous recommendations have been made to improve fire safety in high-rise residential buildings and amendments have been introduced to the RRO to strengthen fire safety responsibilities for employers and responsible persons.

The Building Safety Act 2022 will become a significant piece of legislation for the management of fire safety and will include various provisions for fire safety in high-risk buildings and these are covered in the following chapter.

In addition, the Fire Safety Act 2021 has been introduced, and this Act further qualified the scope of the RRO to include external facades of buildings and entrance doors to individual residential units.

What type of premises does the RRO apply to?

The RRO applies to virtually all premises and covers nearly every type of building, including open spaces.

The order applies to

- offices
- shops
- factories
- care homes
- hospitals
- churches and other places of worship
- village and community halls
- schools, colleges and universities
- pubs, clubs and restaurants
- sports centres
- tents and marquees
- hotels and hostels
- warehouses
- shopping centres.

The RRO does not apply to private dwellings, including individual flats in a block or house, but it does include the common parts of residential buildings and the doors between domestic premises and common parts.

Fire safety requirements for houses in multiple occupation are contained in Housing Act legislation.

Who is legally responsible for complying with the RRO?

The RRO refers to a 'responsible person' as being the person with responsibility for complying with the RRO.

Who is the 'responsible person'?

Article 3 of the RRO defines the 'responsible person' in relation to a workplace as the employer, if the workplace is to any extent under their control.

In relation to premises that are not a workplace, the person who has control of the premises (as occupier or otherwise) in connection with the carrying on by them of a trade, business or other undertaking (for profit or not) is deemed the 'responsible person'.

Where the person in control of the premises does not have control in this way, the 'responsible person' is deemed to be the owner.

What happens if there is more than one person in control of premises?

Where there is more than one responsible person in any type of premises – for example, a multi-occupied complex – all must take reasonable steps to co-operate with one another.

If there are managing agents responsible for common parts, then they will be the responsible person for those areas.

The requirements of the RRO are in fact imposed on any person who has, to any extent, control of premises, so far as the requirements of the RRO relate to matters within their control.

Does the duty to manage fire safety apply just to employers?

No. The RRO does not specifically refer to employers but refers to 'relevant persons'.

Relevant person includes

- any person (including the responsible person) who is or may be lawfully on the premises
- any person in the immediate vicinity of the premises who is at risk from a fire on the premises.

This includes everyone – employees, public, customers, contractors, visitors, students, patients and so on.

The responsible person must consider the fire safety needs of those at special risk – for example, people with disabilities, young people and people not familiar with the building.

The RRO introduces clear responsibility for the 'responsible person' to consider the needs of the public in respect of fire safety and not just employees.

What are the main requirements of the RRO?

The main requirement of the RRO is for the responsible person to carry out a fire risk assessment.

The fire risk assessment must focus on the safety in cases of fire of all 'relevant persons'.

In addition to a fire risk assessment, the responsible person must

- appoint one or more competent persons to assist with fire safety
- ensure that the principles of prevention are applied to any preventative and protective measures implemented for fire safety
- provide employees with clear and relevant information on the risks to them identified by the fire risk assessment, about measures taken to prevent fires and, generally, how they will be protected if a fire occurs
- consult with employees about fire safety matters
- take special steps regarding the employment of young people and inform their parents about the fire measures that have been taken
- inform all persons who are not employees about fire safety measures in the premises – for example, contractors and temporary workers
- co-operate with other responsible persons and coordinate fire safety matters in multi-occupied buildings
- inform the employer of any temporary workers from an outside organisation with clear and relevant information about how the safety of their employees will be protected in the event of fire
- consider any special precautions needed for managing fire safety from any dangerous, flammable or hazardous substances, including explosives
- establish a suitable means for contacting the emergency services and providing them with relevant information about dangerous substances
- provide all relevant persons, where reasonable, with suitable information, instruction and training on fire safety
- provide and maintain suitable firefighting equipment, fire detection equipment, fire warning systems, emergency lighting and so on, as is deemed necessary in the premises by the fire risk assessment
- provide and maintain unobstructed emergency exit routes which lead to a place of safety
- advise employees that they have a duty to co-operate with their employer to enable them to comply with their statutory duties.

The responsible person must appoint a competent person to provide safety assistance. What does this mean?

The responsible person must appoint one or more competent persons to assist them in undertaking the preventive and protective measures required by the RRO.

The competent person must be able to assist the responsible person to

- avoid fire risks
- evaluate the risks that cannot be avoided
- combat the risks at source through fire protection measures and fire precautions
- adapt to technical progress, such as in the use of sprinklers, alarm systems and early warning systems
- replace the dangerous by the non-dangerous or less dangerous – for example, advising on the use of less flammable materials and substances and replacing the use of solvents
- develop a coherent overall prevention policy that covers technology, organisation of work and the influence of factors relating to the working environment
- give collective protective measures priority over individual protective measures
- give appropriate instructions to employees.

Can anyone be appointed as the competent person?

Yes, provided they have enough knowledge and experience to understand the basic principles of fire safety and to be able to implement the measures contained in the RRO.

There are no formal qualifications to become a competent person, but as the onus is on the responsible person to appoint someone – or more than one person if necessary – to provide safety assistance, the responsible person will need to be able to demonstrate that the person so appointed has suitable credentials to take on the role.

The type of person who could be classed as a competent person could be

- fire safety consultants
- fire engineers
- health and safety managers/officers – provided they have fire safety training
- employers – if they have attended basic fire safety training
- unit/department managers with fire safety training
- health and safety consultants.

A new article 9A, introduced by section 156 of the Building Safety Act 2022, will require that a responsible person must not appoint a person to assist them in the undertaking or reviewing of a fire risk assessment unless that person is competent.

> 9A: Risk assessment: assistance.
>
> > (1) The responsible person must not appoint a person to assist them with making or reviewing an assessment under article 9 unless that person is competent.

> (2) A person is to be regarded as competent for the purposes of this article where the person has sufficient training and experience or knowledge and other qualities to enable the person properly to assist in making or reviewing the assessment.
>
> (3) Where the responsible person appoints more than one person, the responsible person must make arrangements for ensuring adequate co-operation between them.

A person is regarded as competent for this purpose where they have sufficient training, experience or knowledge and other qualities to enable them to undertake this role. This duty applies to the appointment of either an individual or a company. Where the person appointed is a company it is envisaged the responsible person would need to establish from the company that its staff were considered competent.

The Fire Sector Federation, formerly the Fire Risk Assessment Competency Council, published a guidance document in 2020, A Guide to Choosing a Competent Fire Risk Assessor, in order that the responsible person could ensure that they appoint a suitable person to assist them in carrying out their legal duties.

What are the principles of prevention in respect of fire safety?

The RRO requires all responsible persons to undertake fire risk assessments of their premises, operations and any other business activity which could expose persons to risks to their safety.

If the risk assessment indicates that preventative and protective measures are required, then those measures must be in accordance with the requirements of Schedule 1, Part 3 of the RRO.

The principles of prevention are

- avoiding risks
- evaluating the risks which cannot be avoided
- combating risks at source
- adapting to technical progress
- replacing the dangerous with the non-dangerous or less dangerous
- developing a coherent overall protection policy which covers technology, organisation of work and the influence of factors relating to the working environment
- giving collective protective measures priority over individual protective measures
- giving appropriate instructions to employees.

The principles of prevention listed above are more detailed than in other health and safety legislation.

Where can I obtain further information on complying with the RRO?

All fire authorities should be able to provide information and advice on complying with fire safety requirements.

However, in order to address the needs of the business community in understanding the duties placed on them by the RRO, the government published a series of fire safety risk assessment guides during the years 2006–2007. These guides set out everything an employer – or responsible person – will need to do to comply with the law and ensure fire safety standards are maintained.

The guides cover

- offices and shops
- factories and warehouses
- sleeping accommodation
- small and medium places of assembly
- large places of assembly
- theatres and cinemas
- educational premises
- residential care premises
- outdoor events
- healthcare premises
- transport premises and facilities
- animal premises and stables
- means of escape for people with disabilities.

All the guides are available for free download on the government website: www.gov.uk. Search for fire safety law and guidance documents for business. Further guides for residential accommodation have been published in 2023. These include 'Fire safety in purpose-built blocks of flats', 'A guide to making your small block of flats safe from fire' and Fire Safety (England) Regulations 2022: Fire Doors.

What is the Fire Safety Act 2021 (FSA)?

The FSA amends the current RRO, clarifying its scope to include the risks posed by the external facade of buildings and individual entrance doors to flats, thereby implementing the recommendations made by Sir Martin Moore-Bick in his Phase 1 report arising out of the Grenfell Tower Inquiry.

In summary, the FSA

- amends the RRO to require all responsible persons (the relevant duty holder under the legislation) to assess, manage and reduce the fire risks posed by the structure and external walls of the buildings for which they are responsible (including cladding, balconies and windows) and individual doors opening onto common parts of the building
- applies to all multi-occupied residential buildings and is not dependent on the height of the building
- allows the fire and rescue service to enforce against non-compliance in relation to the external walls and the individual doors opening onto the common parts of the premises.

The FSA came in to force in stages.

Section 1 comes into force

- in relation to premises in England, on such day as the Secretary of State may by regulations made by statutory instrument appoint

- in relation to premises in Wales, on such day as the Welsh Ministers may by regulations made by statutory instrument appoint
- different days may be appointed for different purposes.

Section 2 came into force

- at the end of the period of two months beginning with the day on which this Act was passed.

Section 3 comes into force

- at the same time as section 1 comes fully into force in relation to premises in England.

The FSA requires the owners and managers of multi-occupied residential buildings to ensure that the fire risk assessments for such buildings are reviewed and updated to encompass the structure, external walls and flat entrance doors.

A failure to comply with the requirements of the FSA in relation to the fire risk assessment for external walls and doors onto communal areas could lead to the fire authority taking formal action, as set out in the RRO.

How do the Building Regulations affect fire safety?

The Building Regulations set down standards to which new and refurbished buildings must conform in terms of construction design and safety.

Fire safety requirements for new building works and renovations, alteration, refurbishments and extension to existing buildings are covered in Part B of the Regulations.

The Building Regulations themselves are goal-setting and are supported by 'approved documents', which set out some practical ways in which to meet the legal requirements.

The fire safety aspects of Part B cover

- means of escape
- fire alarms
- fire spread
- access and facilities for the fire service.

The Building Regulations are enforced by the building control department of the local authority and a failure to comply with the requirements could lead to fines and an enforcement notice.

Building control departments must consult with the fire authority when necessary.

If a building complies with Part B of the Building Regulations, it will, in most cases, comply with fire safety legislation. However, there may be operational matters which are outside the remit of Building Regulations but which the fire authority may require additional controls or standards.

There are a number of premises which are subject to various licensing conditions – for example, residential care homes, hospitals, boarding schools, licensed premises and petrol filling stations. Should this be the case, the building control department is obliged to consult with all interested bodies, including the Licensing Authority, regarding fire safety issues.

What are the proposed amendments to the RRO in the Building Safety Act 2022 (BSA)?

Section 156 introduces a requirement for all responsible persons to record their fire risk assessment in full. It replaces the current requirement to only record significant findings of the assessment in specified circumstances.

At present all responsible persons are required to undertake a fire risk assessment, but they are required to record the significant findings of that assessment only when they are an employer of five or more employees or where there is a licence in place or an alterations notice under the RRO in place requiring this.

A new article 9A will require that a responsible person must not appoint a person to assist them in the undertaking or reviewing of a fire risk assessment unless that person is competent. A person is regarded as competent for this purpose where they have sufficient training, experience or knowledge and other qualities to enable them to undertake this role. This duty applies to the appointment of either an individual or a company. Where the person appointed is a company it is envisaged the responsible person would need to establish from the company that its staff were considered competent.

Where there are two or more appointed persons, the responsible person must make arrangements to ensure adequate cooperation between them.

Article 11(2) will be amended to require all responsible persons to record fire safety arrangements irrespective of number of employees.

At present, responsible persons are required to implement appropriate measures for their premises to enable effective planning, organisation, control, monitoring and review in order to comply with the requirements of the RRO. Currently, recording is required only where they employ five or more employees or where there is a licence in place, or an alterations notice issued under the RRO requires recording. A requirement for all responsible persons to record fire safety arrangements will assist responsible persons to demonstrate compliance and enforcing authorities to evidence a breach of the RRO more easily. This amendment will ensure consistency in approach with article 9A of the RRO, which requires competency to be demonstrated. The responsible person will be able to access guidance to assist their recording of their fire safety arrangements.

A new article 21A will require responsible persons to provide residents of buildings containing two or more sets of domestic premises with specific information about relevant fire safety matters. This includes

- information about any risks to residents of the premises identified within the risk assessment

- preventative and protective measures
- name and UK address of the responsible person
- the identity of any person appointed by the responsible person to assist them with making or reviewing the risk assessment under article 9
- the identity of any competent person nominated by the responsible person to implement firefighting
- fire detection measures
- risks to relevant persons throughout the building identified by other responsible persons in the building.

The Secretary of State will also be given powers to make regulations to extend the fire safety information that responsible persons must provide and to specify the format and frequency for which the information must be provided.

The RRO will also be amended to include provision for amending Article 22 of the Fire Safety Order to require that all responsible persons for premises regulated under the RRO establish whether any other responsible person in the premises shares, or has, duties in respect of the premises and to inform each other of their name, address and parts of the premises they consider themselves to be responsible person for. To inform and support responsible persons in discharging this duty, best practice will be provided in guidance.

A new Article 22A will apply to all premises under the RRO. All departing responsible persons of relevant premises must provide the relevant fire safety information specified in a new Article 22A(3) to incoming responsible persons before they depart. It is intended that this requirement will facilitate the preservation of relevant fire safety information between successive responsible persons across the lifetime of a building. These complement the 'golden thread' provision for accountable persons in the BSA.

A new article 22B will apply to premises consisting of, or including, a residential unit in a higher-risk building (a building in England that (a) is at least 18 m in height or has at least seven storeys and (b) contains at least two dwellings or any units of living accommodation) and require responsible persons for such premises to co-operate with accountable persons in the premises to enable them to carry out their duties under the Building Safety Act. There is a reciprocal duty of co-operation between accountable persons and responsible persons in section 109 of the BSA.

The purpose of this requirement is to ensure that the duties of responsible persons and accountable persons will contribute towards a whole-building approach to building and fire safety.

Fines will become unlimited for failing to comply with requirements relating to the installation of luminous tube signs, failing to comply with any requirements imposed by an inspector and impersonating an inspector.

New guidance documents will be issued by the government and where responsible persons follow the guidance they will be deemed to be complying with their duties under the RRO.

The new guidance documents will clearly define what is 'statutory guidance' (and must be followed), as is currently the case with Health and Safety Executive approved codes of practice.

(NB: for more detail on the BSA please refer to Chapter 3.)

What is included in the Fire Safety (England) Regulations 2022?

The Fire Safety (England) Regulations 2022 are being introduced under article 24 of the RRO to implement the majority of the recommendations made to government in the Grenfell Tower Inquiry Phase 1 report, which require a change in the law.

The regulations came into force on 23 January 2023.

These regulations make it a requirement in law for responsible persons of high-rise blocks of flats to provide information to fire and rescue services to assist them to plan and, if needed, provide an effective operational response.

Also, the regulations require responsible persons in multi-occupied residential buildings that are high-rise buildings, as well as those above 11 m in height, to provide additional safety measures.

In all multi-occupied residential buildings, the regulations require responsible persons to provide residents with fire safety instructions and information on the importance of fire doors. The regulations apply to existing buildings, and requirements for new buildings may be different.

In high-rise residential buildings, responsible persons will be required to

- *building plans*: provide their local fire and rescue service with up-to-date electronic building floor plans and to place a hard copy of these plans, alongside a single page building plan that identifies key firefighting equipment, in a secure information box on site.
- *external wall systems*: provide their local fire and rescue service with information about the design and materials of a high-rise building's external wall system and inform the fire and rescue service of any material changes to these walls. Also, they will be required to provide information in relation to the level of risk that the design and materials of the external wall structure gives rise to and any mitigating steps taken.
- *lifts and other key firefighting equipment*: undertake monthly checks on the operation of lifts intended for use by firefighters and evacuation lifts in their building, and check the functionality of other key pieces of firefighting equipment. They will also be required to report any defective lifts or equipment to their local fire and rescue service as soon as possible after detection if the fault cannot be fixed within 24 h, and to record the outcome of checks and make them available to residents.
- *information boxes*: install and maintain a secure information box in their building. This box must contain the name and contact details of the responsible person and hard copies of the building floor plans.
- *wayfinding signage*: install signage visible in low light or smoky conditions that identifies flat and floor numbers in the stairwells of relevant buildings.

In residential buildings with storeys over 11 m in height, responsible persons will be required to

- *fire doors*: undertake annual checks of flat entrance doors and quarterly checks of all fire doors in the common parts.

In all multi-occupied residential buildings with two or more sets of domestic premises responsible persons will be required to

- *fire safety instructions*: provide relevant fire safety instructions to their residents, which will include instructions on how to report a fire and any other instruction that sets out what a resident must do once a fire has occurred, based on the evacuation strategy for the building.
- *fire door information*: provide residents with information relating to the importance of fire doors in fire safety.

Case studies: Breach of the RRO

Hotel. Radnor Hotel – £200 000 (2015)

In March 2015, the former owner of the Radnor Hotel in Bayswater, London, was fined £200 000 for breaches of the RRO.

When fire safety officers visited the six-storey, 18-room hotel to carry out a routine inspection, they found

- inadequate fire detection systems
- blocked emergency exits
- inadequate emergency lighting
- fire doors missing or tied open with extension cords or string
- the basement room used for sleeping
- no risk assessment had been carried out.

The owner was issued with an enforcement notice. They failed to comply with it.

The hotel continued to operate without a working fire detection system.

They were fined £200 000 plus £30 000 court costs for seven breaches of the RRO.

They were also sentenced to four months' imprisonment, suspended for 18 months.

Supermarket. The Co-operative Group – £210 000 (2010)

The Co-operative Group was charged with serious fire safety breaches at its store in Shirley Road, Southampton. Hampshire Fire and Rescue Authority prosecuted it for six breaches of fire safety under the RRO in 2010.

The fire safety breaches were

- failing to maintain the rear emergency exit doors
- a fitted lock requiring a security code on the emergency door

- fire alarm call point obstruction
- failing to ensure that the store manager was provided with suitable and sufficient fire safety training
- failing to ensure that the fire alarm system was being regularly tested
- failing to ensure a means of early detection of fire.

Three years earlier the Co-Operative Group was fined £250 000 for fire safety offences at several stores in East Sussex.

Office accommodation. Shell International – £300 000 (2009)
Multinational oil and gas company Shell International was fined over significant failings in fire safety at the Shell Centre in central London. The energy company pleaded guilty at Inner London Crown Court to three breaches of the RRO. Two small fires in three weeks at the Shell Centre on York Road, Waterloo, prompted a wide investigation.

The fire safety breaches were

- blocked escape routes
- blocked fire exits
- defective fire doors
- excessive fire loading.

Retail store. New Look – £400 000 (2009)
British global fashion retailer New Look, which has a chain of high street shops in the UK, received a fine of £400 000 following a fire that gutted the retailer's Oxford Street store in 2007. The blaze required 35 engines and 150 firefighters to tackle it, and crews remained at the scene for three days. Trade was disrupted at more than 50 Oxford Street shops. New Look pleaded guilty to two breaches of the RRO following prosecution by the London Fire Brigade.

The fire safety breaches were

- insufficient staff training
- storage blocking escape routes
- lack of suitable and sufficient fire risk assessment.

Residential care home. A group care home business – £432 944.64 (2020)
A residential care home business in Cardiff was fined more than £430 000 for serious fire safety failings, following a prosecution brought by South Wales Fire and Rescue Service.

The business, which runs multiple facilities across the UK including a residential care home in Cardiff, was sentenced on 10 September 2020 at Cardiff Magistrates Court.

South Wales Fire and Rescue Service said that, following multiple inspections dating back to 2011, the 25-bedroom residential care home was found to be in breach of several fire safety regulations within the RRO.

Over several years, extensions of time to comply with remedial measures were granted for the company to rectify the failings but the requirements to comply with the RRO were not met, which resulted in prosecution.

The fire safety breaches were

- lack of fire safety arrangements including building structure fire protection to prevent the spread of fire and smoke
- inadequate fire risk assessment
- insufficient smoke alarms
- inaccessible blocked fire escape routes
- deficient emergency lighting
- failure to conduct appropriate evacuation drills
- lack of maintenance on critical escape routes
- substandard fire safety management.

In summing up, the District Judge is reported to have said

> This is firmly in the category of high culpability . . . these are the most vulnerable members of our society . . . each of them was put at risk . . . there was a risk of a large-scale tragedy. These are very serious offences . . . it's very difficult to imagine a more vulnerable class of individual and two of the offences were committed whilst the company was being investigated.

The District Judge ordered that the company pay a sum of £432 944.64 within 12 months. The figure included surcharge and prosecution costs.

KEY RESOURCES

Regulatory Reform (Fire Safety) Order 2005
Fire Safety Act 2021
Building Safety Act 2022
Fire Safety (England) Regulations 2022
Fire safety guidance documents: www.gov.uk – search for fire safety
Fire safety law and guidance documents for business: www.gov.uk – search for fire safety

Perry P
ISBN 978-0-7277-6647-2
https://doi.org/10.1680/fsbqa.66472.027

Chapter 3
The Building Safety Act

How does the Building Safety Act 2022 (BSA) influence fire safety matters?

The BSA will amend the Regulatory Reform (Fire Safety) Order 2005 (RRO) to require that all responsible persons (RPs) will record their fire risk assessments and fire safety arrangements and appoint only competent persons to undertake those assessments.

RPs will also need to work with other RPs in the same premises. In the case of residential higher-risk buildings, RPs will need to co-operate with the accountable person.

The BSA will also strengthen the role of the principal designer under the Construction (Design and Management) Regulations 2015 (CDM 2015) so that, as well as construction site safety duties, principal designers will now be expected to plan, manage and monitor design work to ensure compliance with building regulations and maintain a 'golden thread' of data on design decisions for fire and structural safety.

Does the BSA apply to all of the UK?

No. The BSA applies to England, although selected provisions will apply to the devolved nations.

It is expected that Wales will adopt the BSA and will introduce specific measures by way of regulations in due course.

Scotland will be taking a different route to improving building safety in respect of high-rise buildings and plans to amend current legislation, especially by way of the building code and regulations framework.

Northern Ireland would appear to be adopting a similar route to Scotland and will be amending existing legislation.

What is the BSA?

The BSA takes forward the government's commitment to fundamental reform of the building safety system. It introduces a new regulatory regime for high-risk residential buildings, now defined as 'higher-risk buildings' and buildings 'in scope' – that is, lower rise buildings that demonstrate a higher fire safety risk.

The Explanatory Notes to the BSA provide a proposed definition of higher-risk buildings as being

> the 'higher-risk buildings', which will be regulated by the new occupation regime, as at least 18 m in height or having at least seven storeys and containing at least two residential units. Residential units are either a dwelling or any other unit of living accommodation, for example a room for a residential purpose where basic amenities, such as cooking facilities, a toilet and personal washing facilities are shared with others in the building.

Houses, flats, serviced apartments, supported accommodation and student accommodation facilities are all expected to be within the scope of higher-risk buildings.

Residential care homes, prisons, hotels, hospitals and hospices are likely to be excluded from the provisions of the new Act but may be included later under separate regulations.

The Secretary of State is authorised to update the list of higher-risk buildings, based on emerging risk evidence.

Part 2 of the BSA establishes a new national Building Safety Regulator, which will sit in the Health and Safety Executive and report to the Secretary of State. The Building Safety Regulator will be responsible for implementing the new regulatory regime, overseeing the registration and inspection of higher-risk buildings, and has express powers to authorise remedial works, stop non-compliant projects, appoint special measures for failing projects and order the replacement of key fire safety officers. The Regulator will establish and consult with three advisory committees on building, industry competence and resident representation while developing the new building safety regime.

Part 3 of the BSA empowers the Secretary of State to impose competency requirements on duty holders working on higher-risk buildings. Those appointing duty holders will also be required to ensure all appointees and prescribed persons meet competency requirements and provide documents demonstrating competency. Full details of duty holder responsibilities and the new Gateway system for approving the design, construction and occupation of in-scope buildings will be covered in secondary legislation.

Part 4 of the Act establishes the role of the 'accountable person' (AP) who will be legally responsible for the safety of higher-risk buildings. APs may be individuals or corporate entities and will hold either a legal estate in possession of the common parts of the building or a relevant repairing obligation in respect of the common parts (meaning that management companies may also be APs).

Further guidance is promised on how APs will be identified.

APs have ongoing obligations to assess and prevent fire safety risks, update prescribed building safety information (formerly referred to as 'the golden thread'), register higher-risk buildings and apply for a building assurance certificate before the building can be occupied.

APs must proactively engage with residents, provide them with key building information and develop a resident engagement strategy.

The BSA updates the Landlord and Tenant Act 1985, placing express duties on residents of higher-risk buildings to keep their property in good repair, co-operate with building safety managers to fulfil building safety duties and not hinder or damage fire safety works or equipment.

APs will have rights to enter residents' dwellings and issue enforcement notices, which shall be enforceable by way of the courts. The Act imports a new regime for charging tenants building safety costs incurred carrying out building safety work. However, landlords will not be able to recharge residents for work that are the subject of enforcement proceedings by the Regulator or as a result of the AP's negligence or breach of contract.

The BSA will impose implied terms on lease agreements covered by the 1985 Act to incorporate this new regime.

The BSA provides details about liability and sanctions for breaches of the building safety regime. Duty holders who fail to meet key building safety obligations will be guilty of criminal offences.

Where a duty holder is a corporate entity, individuals within that organisation may also be prosecuted where the breach was committed with their consent or connivance or because of their neglect.

APs who fail to meet key obligations will also be guilty of a criminal offence, punishable by imprisonment for up to two years and/or a fine.

The BSA also amends the Building Act to allow prosecutions for breaches of the Building Regulations for up to ten years. However, the Act does not anticipate duty holders being placed under statutory duties or allowing individuals to bring actions in civil court against duty holders.

Part 5 of the BSA requires the Secretary of State to establish a scheme for all developers to allow complaints to be made to a newly established New Homes Ombudsman.

The Secretary of State will issue guidance about the adequacy of insurance schemes for higher-risk buildings and building safety works.

The Secretary of State will also establish a new regulatory system for construction products marketed in the UK, identifying 'designated products' and 'safety critical products' that may be withdrawn from the market if they present a risk.

The BSA also amends the Architects Act 1997 to allow the Architects Registration Board to monitor the competence of architects in line with the new regime.

The BSA will also amend the RRO.

What is the role of the Building Safety Regulator (BSR)?

The BSR will improve the safety and standard of buildings through three critical functions, namely

- leading the delivery of the new regulatory regime for high-rise and other buildings in scope
- overseeing the safety and performance of all buildings
- promoting the competence and organisational capability of professionals, tradespeople and building control professionals working on all buildings.

The BSR will have new powers and responsibilities to ensure the safety of all buildings and will have additional responsibilities for higher-risk buildings. These additional powers apply not only to the design, construction and refurbishment of such buildings, but also to their ongoing safety management while they are occupied, so that residents are safe – and feel safe – in their homes.

It will be independent and give expert advice to local regulators, landlords and building owners, the construction and building design industry and to residents.

What are the duties of the Building Safety Regulator?

There are three main duties allocated to the Building Safety Regulator, namely

- to oversee the safety of people in relation to risks arising from buildings and to improve the standards of all buildings
- to directly regulate the safety of higher-risk buildings during construction and after occupation
- to improve the competence of all those involved in the construction and management of higher-risk buildings.

The Building Safety Regulator will be the building control authority for higher-risk buildings.

Other duties and responsibilities of the Building Safety Regulator will be to

- publish a new Code of Conduct for building inspectors
- carry out misconduct investigations
- issue stop notices
- levy unlimited civil fines and commence criminal prosecutions, which attract prison sentences of up to two years
- appoint a special measures manager where there are serious failures on the part of APs or building safety managers.

The Regulator's prosecutorial powers will extend to directors, managers and officers of companies that have committed an offence with their consent or connivance, and will permit the Regulator to extend time limits for prosecution to ten years for certain notices issued under the Building Act 1984.

A new role of 'accountable person' (AP) is to be created under the Building Safety Act. Who is this and what are their duties?

The AP is the entity responsible for meeting the statutory obligations for occupied higher-risk buildings. The AP will usually be the person who either has the legal estate in possession of, or is under a relevant repairing obligation for, any part of the common parts of the building.

Where there are two or more persons with legal interest in a building in scope, the AP will be the one responsible for the repair and maintenance of any common parts of the building.

Where there are two or more APs for a building, the person/organisation responsible for the exterior and structure of the building will be designated 'principal accountable person'.

Example

In a building with a single AP, that AP will be responsible for all the AP duties in Part 4 of the Building Safety Act.

In a building where there is a freeholder who retains repairing obligations for the exterior and structure of the property, and there is a superior leaseholder who retains repairing obligations for all the other 'inner' common parts, both entities will be APs and the freeholder will be the principal accountable person.

In a building where there is a freeholder who does not retain repairing obligations, a superior leaseholder with repairing obligations for the exterior and structure of the property and a management company where every lease in the property states that the management company is responsible for the remainder of the common parts, the superior leaseholder and the management company will be APs. The superior leaseholder will be the principal accountable person.

(Example taken from Government Explanatory Notes on the Building Safety Act July 2021)

The AP may be an individual or a corporate entity such as the building owner or a management company. This is the legal person who under the new regime will be responsible for the fire and structural safety of a 'higher-risk' building.

The AP will be responsible for assessing building safety risks on an ongoing basis and for taking all reasonable steps to manage those risks by preparing a safety case report, which must be submitted to the Building Safety Regulator.

The AP could be the building owner, the managing agents or any individual or corporate entity that has repairing obligations for the building.

New criminal sanctions (including imprisonment up to two years and unlimited fines) will apply to the AP. Criminal liability could also attach to individual directors, managers and members of any body corporate designated as APs, where there is consent, connivance or neglect by the individual.

Guidance will be issued by the government in due course.

An AP's responsibilities include

- registering a building with the Building Safety Regulator
- applying for a building assessment certificate, which is issued when the Building Safety Regulator is satisfied the AP is complying with their duties under the Act
- assessing (and revising as necessary) building safety risks and taking all reasonable steps to prevent the occurrence, and control the impact of, a major incident arising from building safety risks in or around the building
- preparing (and revising as necessary) a safety case report for a building, which contains their assessment of the building safety risks and the steps that have been taken to prevent a major incident
- keeping information on a building and ensuring it is up to date
- developing systems for investigating complaints and for mandatory occurrence reporting.

What is a 'higher-risk building' under the Building Safety Act?

The exact wording in the Act states that

> 'higher-risk building' means a building in England that
>
> (*a*) is at least 18 metres in height or has at least 7 storeys, and
> (*b*) contains at least 2 residential units.

The Secretary of State may make further Regulations supplementing this section.

The definition in the Act of residential unit is

> 'residential unit' means
>
> (*a*) a dwelling, or
> (*b*) any other unit of living accommodation.

Therefore, in addition to residential buildings, care homes and hospitals can also be considered higher-risk buildings if they are higher than 18 m or have at least seven storeys and so would fall under the new regulatory regime.

The Secretary of State has issued new regulations – The Higher-Risk Buildings (Descriptions and Supplementary Provisions) Regulations 2023 – that list exclusions to the general definition of higher-risk building listed above. The regulations state that buildings are not higher-risk buildings if they comprise entirely of

- a care home
- a hospital
- a secure residential institution

- a temporary leisure establishment or
- military barracks.

The Higher-Risk Buildings (Descriptions and Supplementary Provisions) Regulations 2023 came in to force on 6 April 2023. These Regulations determine which buildings will be subject to the new more stringent safety regime established by the Building Safety Act 2022 ('the 2022 Act'). They do so by supplementing and completing the definitions of 'higher-risk building' at section 120D of the Building Act 1984 ('the 1984 Act') and section 65 of the 2022 Act. The 1984 Act makes provision about the design of, and building work carried out in respect of, 'higher-risk buildings'. The 2022 Act makes provisions for safety management in occupied 'higher-risk buildings'.

How is a storey defined or measured?

For the purposes of section 120D(2)(a) of the Building Act 1984 and section [62](1)(a) of the 2022 Act, when determining the number of storeys a building has

- any storey which is below ground level is to be ignored
- any mezzanine floor is a storey if its internal floor area is at least 50% of the internal floor area of the largest storey in the building which is not below ground level.

For the purpose of the regulations, a storey is treated as below ground level if any part of the finished surface of the ceiling of the storey is below the ground level immediately adjacent to that part of the building.

What are building safety risks as defined in the BSA?

A 'building safety risk' is defined as any risk to the safety of persons in or about a building arising from fire, structural failure or any other matter prescribed in regulations.

The clause is drafted to capture risks inside and outside the building, such as fire spreading on the external wall of a building, but is limited to risks that are associated with the building and/or arising from the building itself. Risks are also limited to those affecting people in and about the building.

The explanatory notes to the Act state that the scope is only intended to encompass major and rapid onset events and that the government 'has no current plans for the higher-risk building regime to extend beyond the risks of fire and structural integrity.' However, the Secretary of State can request advice from the BSR about new and emerging risks and may introduce additional definitions for building risk.

Statutory guidance may also be issued to help determine building risk and may include electrical and gas failures as a building safety risk and integrity of windows and glazing due to high winds and so on.

Draft Regulations have been published – The Higher-Risk Buildings (Management of Safety Risks etc) (England) Regulations 2023 – that give clarity to how building risk is to be interpreted.

The principles prescribed for the purposes of section [84(3)] of the Building Safety Act 2022 are to

- avoid building safety risk
- evaluate building safety risks that cannot be avoided, including identifying the proportionate measures required to reduce, address and mitigate the building safety risks
- combat building safety risks at source by introducing proportionate measures to reduce, address and mitigate that risk at the earliest opportunity
- ensure suitable and proportionate systems are in place for the effective inspection, testing and maintenance of the efficacy of measures taken
- give collective protective measures priority over individual protective measures
- adapt to technical progress
- where reasonable to do so, replace the dangerous with the non-dangerous or less dangerous
- consider the impacts on residents and carry out engagement with residents
- give appropriate instructions and information to employees and persons working on or in the building
- ensure that any person responsible for or assisting with compliance with a duty in or under Part 4 of the 2022 Act has the relevant competence.

Who is responsible for managing the building risks associated with a higher-risk building?

The accountable person (AP) is responsible.

Every AP is under an ongoing obligation to carry out an assessment of the building safety risks in/of/arising from and relevant to the part of the higher-risk building for which they are responsible. They must take all reasonable steps and actions necessary to manage the building safety risks, preventing them from materialising in their building, and to mitigate the severity of such an incident if it does occur.

What is the building safety case?

The safety case report should refer to the totality of the building safety information and include all supporting evidence identifying how fire and structural risks are being managed by the building safety manager. This should be followed by a written explanation and justification of the approach being taken to manage risks, referencing the supporting evidence in the safety case.

The building safety case is submitted by the AP to the Building Safety Regulator. In submitting a safety case report for review, the accountable person will make a case to the Building Safety Regulator that the fire and structural safety risks in the building have been assessed, and adequate safety measures have been identified and put in place to manage the risks appropriately.

Key to the process of putting the safety case together will be the established risk assessment principles of

- identifying the hazards
- deciding who might be harmed and how

- evaluating the risks associated with these hazards
- deciding on the necessary control and mitigation measures
- recording those findings and implementing them
- evaluating and monitoring on an ongoing basis.

The building safety case report will include all relevant information on the structure and materials used in the building and the strategies adopted during the design and construction of the building to minimise risks from catastrophic events.

The building safety case report will include a comprehensive fire risk assessment report and emergency plan as required by the RRO.

The Building Safety Regulator will assess the building safety case and will either accept the findings in the report or will require amendments and improvements to the safety regime proposed.

What is the 'golden thread' of information?

Accurate and up-to-date information about the design, construction and ongoing maintenance of buildings is crucial to ensuring building safety. The golden thread, which will operate from design stage through to occupation of buildings in scope, is intended to encourage openness, transparency and accountability.

The intention of the clauses in the Act is to ensure that the right people have the right information at the right time to ensure buildings are safe and building safety risks are managed throughout the building's life cycle.

This information will be held digitally and will ensure that the original design intent and any subsequent changes to the building are captured, preserved and used to support safety improvements.

It is recommended that building information modelling (BIM) is used to collate and share the information.

CDM 2015 requires that a health and safety file is produced by the principal designer for all projects where there is more than one contractor. The health and safety file will contain information relevant to health and safety and can be provided in either digital or hard copy formats. Ideally, the health and safety file can be expanded to include all the relevant fire safety information about the building project.

Certain information from the golden thread will need to be submitted to the Building Safety Regulator at the beginning of the project before planning permission is granted, namely

- full plans, including a detailed specification in relation to how fire and structural safety risks will be managed

- a 3D digital model of the building 'as planned', which must also show the construction products to be used
- a fire statement, addressing fire service access and water availability. This will ultimately be developed into a fire and emergency file, building on the fire statement and updated throughout the construction phase
- a construction control plan setting out how compliance with the Building Regulations will be achieved and how changes will be controlled and recorded.

Major changes affecting safety will need to be agreed with the principal designer and client and approved by the Building Safety Regulator in advance.

Additional information, the building safety case, will need to be submitted to the Building Safety Regulator before the building can be released for occupancy, namely

- handover of building safety information on the final 'as-built' building
- safety case for the building identifying hazards, describing how risks are controlled and detailing safety management systems, including emergency response procedures
- resident engagement strategy setting out how information will be provided to residents
- complaints procedures and wider engagement initiatives.

The safety case must be reviewed and re-submitted to the regulator every five years and also if the building undergoes major refurbishment.

Under the new regime, a building safety certificate will be required before a building can be occupied. The regulator will not grant a building safety certificate unless all the relevant information has been submitted and approved.

What is the building assessment certificate (BAC)?

The BAC is a legal document issued by the Building Safety Regulator that confirms the building has been constructed in accordance with all the statutory provisions and that procedures are in place to manage all major risks which could affect fire safety, structural integrity and any other prescribed matter for the building.

It will be issued when all the relevant information (the building safety case) has been submitted by the AP.

The Building Safety Regulator can issue the BAC with conditions.

Residents will not be permitted to occupy a new development without an official BAC.

A copy of the certificate will need to be displayed in all common parts of the building so that residents know that the building has been designed with safety in mind and have knowledge of the names and contact details of the accountable person and the building safety manager.

What are the 'gateways' referred to in the BSA?

The BSA talks about three gateways during the planning, design and construction phases of the building.

Each of these gateways would have certain requirements that would need to be met through the co-operation of various duty holders.

The first is the planning gateway that focuses on the planning permission process. Gateway two must be met before the construction begins, whereas the third gateway needs to be met before the occupation of the building.

A similar process has been proposed for major building refurbishments and for buildings that switch from commercial to residential usage. At each gateway, the golden thread of information will need to be updated to get to the next stage.

Gateway one: before planning permission is granted

The first gateway is the least onerous of the three new gateways. Its purpose is to ensure that there is consultation with the local fire authority at the earliest possible stage.

Gateway one will apply only to new high-rise residential buildings of 30 m or more.

It may also be extended to other developments in the 'near vicinity' of these buildings, depending on the outcome of the consultation.

The main features of gateway one are

- making local fire and rescue authorities 'statutory consultees' during the planning process, requiring the local planning authority to consult with them before granting planning permission
- requiring the developer client to submit a fire statement covering fire service vehicle access and access to water supplies with the planning application.

Gateway two: before construction begins

Gateway two will apply to new multi-occupied buildings of 18 m height or more and to the significant refurbishment of existing buildings in scope of the Act. The new requirements supplement the current full plans building application stage under the 2010 Building Regulations, and encourage very early dialogue between the regulator and duty holder about risks and how they are being managed.

Duty holders will need to prepare and submit a number of pieces of the 'golden thread' of information about the building before the construction phase is started, namely

- full plans, including a detailed specification in relation to how fire and structural safety risks will be managed
- a 3D digital model of the building as planned, showing the construction products to be used

- a fire and emergency file, building on the fire statement from gateway one and updated throughout construction
- a construction control plan setting out how compliance with the Building Regulations will be achieved and how changes will be controlled and recorded.

Major changes affecting safety will need to be agreed with the principal designer and client and approved by the Building Safety Regulator in advance.

A phased approach will be possible for complex development, which would result in multiple gateway two approvals for different parts of the development.

An inspection regime must also be agreed with the new regulator as part of gateway two. The regulator will have the power to require demolition of non-compliant building work or for that work to be laid open for inspection for compliance.

Gateway three: before occupation

The third gateway is designed to ensure an adequate handover of the building from those involved in its construction to those who will be responsible for managing and maintaining it during occupation.

Gateway three will be introduced on a phased basis for all existing buildings within the scope of the new regulatory regime. Existing building owners/managers will need to carry out significant work to recreate the golden thread of information for these buildings.

The following will need to be in place before gateway three approval will be granted

- a declaration from the principal contractor and the principal designer confirming the building complies with the Building Regulations
- handover of building safety information on the as-built building plans
- registration of the building
- registration of the accountable person responsible for building safety and the appointed building safety manager
- a safety case for the building identifying hazards, describing how risks are controlled and detailing safety management systems, including emergency response procedures.

Once the regulator is satisfied, it will issue a building safety certificate permitting occupation. The regulator may attach conditions to the certificate. The certificate and safety case will be subject to review every five years.

What is mandatory occurrence reporting under the new BSA?

The principal accountable person has a duty to establish and operate an effective mandatory occurrence reporting system and a further duty is placed on accountable persons to ensure that reportable information related to the safety of an occupied building is given to the Building

Safety Regulator in the manner specified by them. The adopted system will be assessed as part of the building assessment certificate application.

Basically, the principal accountable person must have in place a system to report particularly dangerous safety-related events to the Building Safety Regulator.

As an example, an outbreak offire in a refuse chute/room that fails to shut down the ventilation system or automatically close fire doors due to faults must be reported as serious consequences to safety could ensue.

The Building Safety Regulator will impose remedial actions and may check that works have been carried out.

Information has to be kept in relation to higher-risk buildings. What does this mean?

The accountable person (AP) must keep and maintain certain required information about their buildings and ensure that it is kept as up to date as possible. This information is part of the 'golden thread' of information.

The BSA does not define exactly what will be prescribed information, as this will be set out in additional regulations.

Information could include

- details of the fire alarm system
- means of escape details
- fire stopping information
- the fire risk assessment
- emergency plans for people with disabilities
- maintenance schedules for all fire protection and preventative measures
- details of local Fire and Rescue Authority
- schedule of building materials used within the building.

The AP will be under an obligation to share the 'prescribed information' with others who may need to know about the building – for example, contractors who are carrying out refurbishment work, residents and residents groups, the Building Safety Regulator and the local fire and rescue service.

Who will be duty holders under the new Act?

Duty holders will be

- the accountable person or principal accountable person
- the client
- the principal designer

- the principal contractor
- designers
- contractors.

The duties of the accountable person and the building safety manager have been summarised earlier in this guide.

The duties of the client will be

- *culture*: establishing a culture of compliance on the project
- *project arrangements*: making suitable arrangements for managing the building work to deliver compliance with the building regulations, including the allocation of sufficient time, resources and prioritisation
- *key appointments*: appointing a competent principal designer and principal contractor, with named individuals registered as competent with the new building safety regulator
- *compliance*: taking reasonable steps to ensure that the principal designer and principal contractor comply with their building safety obligations
- *golden thread*: establishing appropriate information management arrangements for the building
- *gateways*: ensuring that regulatory requirements to get approval from the new building safety regulator at various stages of the project are met
- *handover*: ensuring an appropriate pre-occupation handover takes place.

The duties of the principal designer will be

- *leading the design:* planning, monitoring and managing the pre-construction phase and coordinating building safety to ensure that the project complies with building regulations
- *monitoring designers:* taking reasonable steps to ensure that all designers are discharging their duties and promoting the statutory building safety objective
- *golden thread:* developing and/or contributing to the information that must be submitted at regulatory gateway points; developing and maintaining a complete golden thread of information and contributing to the handover process prior to occupation
- *compliance:* signing a declaration, along with the principal contractor, that the building complies with the Building Regulations.

The duties under the Act will build on those required by CDM 2015. The principal designer appointed under CDM 2015 will be assumed to be the principal designer under the BSA.

The duties of the principal contractor will be

- *leading construction*: planning, monitoring and managing the construction phase and coordinating building safety to ensure that the project complies with the Building Regulations

- *monitoring the supply chain*: taking reasonable steps to ensure that contractors are meeting their duties and promoting the statutory building safety objective
- *golden thread*: developing and implementing the Construction Control Plan to ensure compliance with full plans and contributing to other information and record-keeping requirements
- *compliance*: obtaining sign-off from the client and principal designer and/or regulator approval before deviating from approved plans during construction; signing a declaration that the building complies with the Building Regulations and that an appropriate handover back to the client has taken place.

As with the principal designer, the new duties of the principal contractor align with those allocated under CDM 2015. The principal contractor appointed under CDM 2015 will be assumed to be the principal contractor under the BSA.

The duties of designers, contractors and others involved with the project will include

- *statutory objective*: an over-arching duty to promote building safety and the safety of persons in and around the building
- *mandatory occurrence reporting regime*: complying with reporting requirements using the system established by the client
- *competence*: only making and accepting appointments where the necessary skills, knowledge, experience and organisational capability are in evidence
- *working together*: co-operating, coordinating, sharing information and liaising with other duty holders and the regulator.

Similar duties apply to these roles under CDM 2015. The BSA and CDM 2015 are expected to work hand in hand so that all aspects of construction site safety and the future occupancy of the building are considered and hazards and risks eliminated or reduced to acceptable standards. Information is critical and must be handed over to those that require it.

The BSA includes provision for residents of existing higher-risk buildings to be included in the decision-making process and to be consulted where appropriate.

What are the key amendments proposed for the Building Act 1984 and the Building Regulations 2010?

It is proposed that the Building Safety Regulator establish and maintain a register of building inspectors (individuals) and building control approvers (either organisations or individuals). Individuals and organisations currently known as 'approved inspectors' who wish to continue undertaking building control work will need to register as 'building control approvers'. The role of a registered building inspector being introduced in this Act is new. A registered building inspector is an individual who will be able to provide advice to local authorities or registered building control approvers overseeing building work. Many inspectors in local authorities and approved inspectors are expected to transition to this role.

Other changes in the BSA include

- removing the ability of a person carrying out building work on higher-risk buildings to choose their own building control body where building control is required
- changes to procedures for initial notices
- allowing the Secretary of State to make regulations as to how the new regulatory regime for higher-risk buildings will be applied (if at all) to a public body designated under sections 5 or 54 of the Building Act 1984
- providing information-gathering powers on building control for the regulator and local authority building control
- requiring the Building Safety Regulator to set up the new national electronic register/ portal of information governing the work of registered building control approvers
- amending the existing powers to give the Secretary of State the ability to designate bodies for the purposes of publishing criteria for and/or approving an insurance scheme
- a 15-year limitation period will apply to section 38 of the Building Act 1984 (essentially, applicable where there is a breach of the Building Regulations).

What will change regarding the competency of architects?

Architects will have to demonstrate relevant competency across all building disciplines. Disciplinary orders made against architects by the Professional Conduct Committee (PCC) of the Architect Registration Board (the Board), the statutory regulator of architects in the UK, must be listed alongside an architect's entry in the architects register. This is to provide greater transparency to members of the public procuring architectural services as the register entry, accessible through the Board's website, will show whether an architect has been disciplined recently. The Board will determine rules for the length of time that a disciplinary order will be listed on the register, depending on the severity of the order.

Architects will also have to demonstrate continuing professional development in order to maintain their entry in the architects register.

What is the residents' panel?

Section 11 requires the Building Safety Regulator to establish a committee known as the residents' panel. The purpose of the panel is to ensure that residents have a voice in the work of the Building Safety Regulator, and the Building Safety Regulator has a broad power to consult the residents' panel on any of its functions that impact the residents of higher-risk buildings.

The panel must include residents, and may also have members from the following groups: owners of residential units (e.g. flats) in higher-risk buildings who are not occupying the property (who could, for example, be impacted by expenses relating to the higher-risk building) and groups or organisations representing residents and/or non-occupying leaseholder owners, advocating for them or supporting them.

The Act also requires that the Building Safety Regulator must consult the residents' panel on certain matters which it is expected would be of particular interest and importance for residents of higher-risk buildings, specifically

- the residents' panel must be consulted on guidance to residents on their rights and obligations under the new regulatory regime for higher-risk buildings in occupation
- the residents' panel must be consulted on guidance relating to any duties under section 91 (provision of information etc. to the regulator, residents and other persons) where the guidance is about information provision to residents
- the residents' panel must be consulted on guidance relating to section 93 (residents' engagement strategy), section 94 (requests for further information), section 95 (complaints procedure operated by the principal accountable person), section 97 (duties on residents and owners) or regulations made under any of those clauses
- section 17 requires that the Building Safety Regulator must consult the residents' panel on its strategic plan
- section 96 requires that the regulator consult the residents' panel on its system for dealing with complaints from residents that are escalated to the Building Safety Regulator.

Will officers, managers and employees of a corporate body be liable for prosecution should offences be committed?

Section 138 provides that, where a corporate body commits a criminal offence under Parts 2 or 4 of this Act, any director, manager, secretary or other similar officer of that body is also deemed to have committed that offence in certain circumstances. Those circumstances are where the individual has consented to or connived in the commission of the offence or where the offence is attributable to any neglect on their part. Section 39 makes similar provision in respect of the criminal offences in the Building Act 1984.

Offences under the Corporate Manslaughter and Corporate Homicide Act 2007 may also apply should fatalities occur as a result of criminal negligence.

KEY RESOURCES

Building Safety Act 2022

Information relating to the Building Safety Act 2022: Department for Levelling Up, Housing and Communities

Information on the Building Safety Regulator: Health and Safety Executive

TOP TIPS

- review all fire safety procedures and ensure that information is made available to all who will need it and that it is shared amongst other duty holders
- appoint only competent fire risk assessors
- know the definition of higher-risk buildings and whether the Building Safety Act applies
- appoint an accountable person where necessary. Put the appointment in writing
- ensure that the accountable person is always in post
- appoint a building safety manager, where applicable. Ensure that they are competent and that the post is always filled
- make appointments in writing and specify responsibilities
- establish ways of introducing the 'golden thread' of information for all information relevant to the safety of the building
- communicate with residents if you own or manage residential buildings, especially about fire safety matters, and share information
- communicate with all building occupiers and others who are resoponsible for parts of the building and share information
- be proactive and take control of your duties under the new legislation. Seek competent advice where necessary
- review the construction materials specified and ensure that they are fit for purpose and meet specifications
- check building and refurbishment works for compliance with the tender or contract documents
- do not turn a blind eye to managing the building works or the building once occupied. Legal duties cannot be delegated
- manage refurbishment works closely and appoint a competent design and project management team.

Perry P
ISBN 978-0-7277-6647-2
https://doi.org/10.1680/fsbqa.66472.045

Chapter 4
Fire risk assessments

What is a fire risk assessment?

A fire risk assessment is, in effect, an audit of your workplace and work activities in order to establish how likely a fire is to start, where it would start, how severe it might be, who it would affect and how people would get out of the building in an emergency.

A fire risk assessment should be concerned with *life safety* and not with matters that are fire engineering matters, although structural elements of the building and materials used in its construction will have a bearing on the likelihood of fire and the rapidity of its spread and should therefore be considered in the fire risk assessment.

A fire risk assessment is a structured way of looking at the hazards and risks associated with fire and the products of fire – for example, smoke.

Like all risk assessments, a fire risk assessment follows *five key steps*, namely

- identify the hazards
- identify the people and the location of people at significant risk from a fire
- evaluate the risks
- record findings and actions taken
- keep assessment under review.

So, a fire risk assessment is a record that shows you have assessed the likelihood of a fire occurring in your workplace, identified who could be harmed and how and decided on what steps you need to take to reduce the likelihood of a fire (and therefore its harmful consequences) occurring. You have recorded all these findings regarding your premises into a particular format called a risk assessment.

Does the fire risk assessment need to be in writing?

Under the Regulatory Reform (Fire Safety) Order 2005 a fire risk assessment does not need to be in writing if there are fewer than five employees in the workplace.

If you have fewer than five employees but a licence for the premises – for example, Premises Licence – then you will need to have a written record of your fire risk assessment.

Where there are five or more employees the fire risk assessment must be in writing.

Changes introduced to the RRO by the Building Safety Act 2022 require that all fire risk assessments are in writing and that all findings shall be recorded and not just those that are deemed significant, as follows:

> In article 9 (risk assessment)—
>
> (*a*) in paragraph (6) for the words from 'record' to the end substitute 'make a record of the assessment or review, which must in particular include the information prescribed by paragraph (7).';
>
> (*b*) in paragraph (7)(a) omit 'significant'.

Definitions

- *Risk assessment*: the overall process of estimating the magnitude of risk and deciding whether or not the risk is tolerable or acceptable
- *Risk*: the combination of the likelihood and consequence of a specified hazardous event occurring
- *Hazard*: a source or a situation with a potential to harm in terms of human injury or ill health, damage to property, damage to the environment or a combination of these
- *Hazard identification*: the process of recognising that a hazard exists and defining its characteristics.

What are the FIVE STEPS of a fire risk assessment?

Step 1. Identify the hazards

Sources of ignition

You can identify the sources of ignition in your premises by looking for possible sources of heat that could get hot enough to ignite the material in the vicinity.

Such sources of heat/ignition could be

- naked flames – for example, candles, fires and blow lamps and so on
- electrical, gas or oil fired heaters
- hot work processes – for example, welding and gas cutting
- cooking, especially frying
- faulty or misused electrical appliances including plugs and extension leads
- lighting equipment, especially halogen lamps
- hot surfaces and obstructions of ventilation grille – for example, radiators
- poorly maintained equipment that causes friction or sparks
- static electricity
- arson

- smokers' materials (look particularly where these could be hidden as people may be smoking furtively in non-designated smoking areas).

Look out for any evidence that things or items have suffered scorching or overheating – for example, blackened plugs and sockets, burn marks, cigarette burns and scorch marks and so on.

Check each area of the premises systematically

- customer area, public areas and reception areas
- work areas and offices
- staff kitchen and staff rooms
- store rooms and cleaners' stores
- plant rooms and motor rooms
- refuse areas
- external areas.

Sources of fuel

Anything (generally) that burns is fuel for a fire. Fuel can also be invisible in the form of vapours and fumes and so on given off from other less flammable materials.

Look for anything in the premises that is in sufficient quantity to burn reasonably easily or could cause a fire to spread to more easily accessible fuels.

Fuels to look out for are

- wood, paper and cardboard
- flammable chemicals – for example, cleaning materials
- flammable liquids – for example, cleaning substances and liquid petroleum gas
- flammable liquids and solvents – for example, white spirit, petrol and methylated spirit
- paints, varnishes, thinners and so on
- furniture, fixtures and fittings
- textiles
- ceiling tiles and polystyrene products
- flammable insulation materials
- waste materials and general rubbish
- gases.

Identifying sources of oxygen

Oxygen is all around us in the air that we breathe. Sometimes, other sources of oxygen are present that accelerate the speed at which a fire ignites – for example, oxygen cylinders for welding.

The more turbulent the air the more likely the spread of fire will be – for example, opening doors brings a 'whoosh' of air into a room and the fire is fanned and intensifies. Mechanical ventilation also moves air around in greater volumes and more quickly.

Do not forget that while ventilation systems move oxygen around at greater volumes, they will also transport smoke and toxic fumes around the building.

Identifying how fire might spread

Consider the construction of the premises. Are there any materials used which would burn more easily than other types? Hardboard, chipboard and blockboard burn more easily than plasterboard.

Thermal insulation may be combustible, so checks on manufacturers' safety data sheets and technical specifications should be made.

External cladding used on many buildings has been identified as a potential fuel for fire spread. Details of the material, its specification and the method of application must be considered as part of the assessment of potential fire hazards.

How might a fire spread from one area to another? Is there adequate fire stopping to the structure, has compartmentation been designed into the building to limit fire spread and could any ventilation systems fuel the flames, cause backdraught or cause smoke to spread?

Do not just think about flames of the fire – smoke travels very fast and is often the cause of people being overcome in a fire and failing to evacuate the building.

Step 2. Identify who could be harmed

You need to identify who will be at risk from a fire and where they will be when a fire starts. The law requires you to ensure the safety of your staff and others – for example, customers. Would anyone be affected by a fire in an area that is isolated? Could everyone respond to an alarm and evacuate?

Will you have people with disabilities in the premises – for example, wheelchair users or people with visual or hearing impairments? Will they be at any greater risk of being harmed by a fire than other people?

Will contractors working in plant rooms, on the roof and so on be adversely affected by a fire? Could they be trapped or not hear alarms?

Who might be affected by smoke travelling through the building? Smoke often contains toxic fumes.

Step 3. Evaluate the risks arising from the hazards

What will happen if there is a fire? Does it matter whether it is a minor or major fire? Remember small fires can grow rapidly to infernos.

A fire is often likely to start because

- people get careless with cigarettes and matches
- people purposely set light to things
- cooking canopies catch fire due to grease build-up
- people put combustible material near flames/ignition sources
- equipment is faulty because it is not maintained
- electrical sockets are overloaded.

Will people die in a fire from

- flames
- heat
- smoke
- toxic fumes.

Will people get trapped in the building?

Will people know that there is a fire, and will they be able to get out?

Step 3 of the risk assessment is about looking at what *control measures* you have in place to help control the risk or reduce the risk of harm from a fire.

Remember – fire safety is about *life safety*. Get people out fast and protect their lives. Property is always replaceable.

You'll need to record on your fire risk assessment the fire precautions you have in place, that is

- what emergency exits do you have and are they adequate and in the correct place
- are they easily identified and unobstructed
- is there firefighting equipment
- how is the fire alarm raised
- where do people go when they leave the building and is there an assembly point
- are the signs for fire safety adequate
- who will check the building and take charge of an incident – that is, do you have a fire warden appointed
- are fire doors kept closed
- are ignition sources controlled and fuel sources managed
- do you have procedures to manage contractors (remember Windsor Castle went up in flames because a contractor used a blow torch near the curtains!)?

Of all your fire safety precautions for the premises, is there anything more that you need to do?

Are staff trained in what to do in an emergency? Can they use fire extinguishers? Do you have fire drills? Is equipment serviced and checked – for example, emergency lights and fire alarms?

Step 4. Record findings and action taken

Complete a fire risk assessment form and keep it safe.

Make sure that you share the information with staff and others that may need the information.

If contractors come to site, make sure that you discuss their fire safety plans with them and that you tell them what your fire precaution procedures are.

Step 5. Keep assessment under review

A fire risk assessment needs to be reviewed regularly – about every six months or so and whenever something has changed, including layout, new employees, new procedures, new legislation, increased stock and so on.

What is the definition of 'responsible person'?

Article 3 of the RRO 2005 states

> In this Order 'responsible person' means—
>
> (*a*) in relation to a workplace, the employer, if the workplace is to any extent under his control;
>
> (*b*) in relation to any premises not falling within paragraph (a)—
>
> (*i*) the person who has control of the premises (as occupier or otherwise) in connection with the carrying on by him of a trade, business or other undertaking
> (for profit or not); or
>
> (*ii*) the owner, where the person in control of the premises does not have control in connection with the carrying on by that person of a trade, business or other undertaking.

The employer is the responsible person for ensuring fire safety within the workplace setting and in any area of the building or grounds under their control.

In a multi-occupied building, such as a tower block let into separate floors or business suites, there will be several responsible persons and each business will need to appoint one.

In a multi-occupied building there will be areas that do not form part of a working space, such as stairways, corridors, plant rooms, entrance lobbies and so on. These areas may still pose a fire safety risk, and someone will need to be responsible for ensuring that fire safety standards are met. The landlord or managing agent will usually be responsible for the maintenance of these areas, and they are deemed to be the responsible person for these areas under the RRO.

Is there any guidance on assessing the risk rating of premises in respect of fire safety?

When completing fire risk assessments, it is sensible to categorise *residual risk* for the premises into a risk rating category – normally referred to as high, medium or low.

In terms of fire risk rating, it is usual to refer to medium risk as 'normal'.

The fire safety guides published to support the RRO also give some advice on risk rating premises.

High risk premises

- any premises where highly flammable or explosive substances are stored or used (other than very small quantities)
- any premises where the structural elements present are unsatisfactory in respect of fire safety
    - lack of fire resisting separation
    - vertical or horizontal openings through which fire, heat and smoke can spread
    - long and complex escape routes created by extensive sub-division of floors by partitions and so on
    - complex escape routes created by the positioning of shop unit displays, machinery and so on
    - large areas of smoke- or flame-producing furnishings and surface materials, especially on walls and ceilings
- permanent or temporary work activities which have the potential for fires to start and spread – for example
    - workshops using highly flammable materials and substances
    - paint spraying
    - activities using naked flames – for example, blow torches, welding
    - large kitchens in work canteens, restaurants
    - refuse chambers and waste disposal areas
    - areas containing foam or foam plastic upholstery and furniture
- where there is significant risk to life in case of fire
    - sleeping accommodation provided for staff, guests, visitors and so on in significant numbers
    - treatment of care where occupants have to rely on others to help them
    - high proportions of elderly or infirm
    - large numbers of people with disabilities
    - where people work in remote areas – for example, plant rooms, roof areas and so on
    - large numbers of people resorting to the premises relative to its size – for example, sales at retail shops
    - large numbers of people resorting to the premises where the number of people available to assist is limited – for example, entertainment events, banquets and so on.

Normal risk premises

- where an outbreak of fire is likely to remain contained to localised areas or is likely to spread only slowly, allowing people to escape to a place of safety
- where the number of people in the premises is small and they are likely to escape via well-defined means of escape to a place of safety without assistance
- where the premises has an automatic warning system or an effective automatic fire fighting system/fire extinguishing/fire suppression system which may reduce the risk categorisation from high.

Low risk premises

- where there is minimal risk to people's lives and where the risk of fire occurring is low, or the potential for fire, heat or smoke spreading is negligible.

Who must I consider when preparing my risk assessment?

Responsible persons must consider the following people as being at risk in the event of a fire

- employees and those on temporary or agency contracts
- employees whose mobility, sight or hearing might be impaired
- employees with learning difficulties or mental illness
- other persons in the premises if the premises are multi-occupied
- anyone occupying remote areas of the premises
- visitors and members of the public, including contractors and so on
- anyone who may sleep on the premises
- anyone with any special needs or disabilities
- residents in higher-risk buildings.

Does a fire risk assessment have to consider members of the public?

A fire risk assessment must be carried out by the responsible person and must consider the risks to the safety of *'relevant persons'* – that is, all persons who are, or could be, lawfully on the premises. This will include members of the public.

The RRO has made the inclusion of all persons, including the public, a legal requirement when completing risk assessments.

Does the fire risk assessment have to include residents?

The RRO does not apply to single private dwellings and so a fire risk assessment does not need to be completed for individual private dwellings and their residents.

All residential buildings that have common parts, communal meeting rooms, laundries, refuse areas and so on must be included in the fire risk assessment, and residents using any communal areas must be included as persons who could be affected by fire.

High-rise residential buildings are subject to increased fire risk assessment scrutiny under the Building Safety Act 2022.

Houses in multiple occupation, care homes, hospitals, supported living accommodation, student accommodation, hotels, guest houses and all other shared residential accommodation must be included in any fire risk assessment and the effects of fire on residents or occupiers must be considered.

The Fire Safety Act 2021 requires that residents be considered from any effect of fire that may occur in common parts and how such a fire may affect their own private dwelling – for example, the ability of flat/apartment front or main access doors to resist the impact of fire.

Who can carry out a fire risk assessment?

The RRO states that the person who carries out a fire risk assessment shall be *competent* to do so.

Competency is not defined specifically in the RRO but is generally taken to mean having a level of knowledge and experience that is relevant to the task in hand.

A fire risk assessment is a logical, practical review of the likelihood of a fire starting in the premises and the consequences of such a fire. Someone who has good knowledge of the work activities and the layout of the building, together with some knowledge of what causes a fire, would be best placed to carry out a fire risk assessment.

The Building Safety Act 2022 has strengthened the requirement for competency in the RRO by inserting a new article 9A.

A new article 9A will require that a responsible person must not appoint a person to assist them in the undertaking or reviewing of a fire risk assessment unless that person is competent. A person is regarded as competent for this purpose where they have sufficient training, experience or knowledge and other qualities to enable them to undertake this role. This duty applies to the appointment of either an individual or a company. Where the person appointed is a company it is envisaged the responsible person would need to establish from the company that its staff were considered competent.

New article 9A: risk assessment: assistance

> (1) The responsible person must not appoint a person to assist them with making or reviewing an assessment under article 9 unless that person is competent.
>
> (2) A person is to be regarded as competent for the purposes of this article where the person has sufficient training and experience or knowledge and other qualities to enable the person properly to assist in making or reviewing the assessment.
>
> (3) Where the responsible person appoints more than one person, the responsible person must make arrangements for ensuring adequate co-operation between them.

Government intends to issue guidance on the topic of fire risk assessor competency and accepts that responsible persons will be able to determine the specific level of competency required dependent on the complexity and risk of the building.

As a result of the Grenfell Tower fire public inquiry the Fire Sector Federation has produced a guide to assessing competency: Approved Code of Practice – A National Framework for Fire Risk Assessor Competency Version 1 – November 2020.

This will help responsible persons to carry out the appropriate checks on individuals and companies.

Can I carry out the fire risk assessment myself?

The duty holder in small or low-risk premises may be able to carry out their own fire risk assessment using guidance available on the local fire authority or government websites and guidance documents issued by government.

The Fire Sector Federation guide states

> duty holders – particularly in small or low-risk premises – may be able to carry out their own fire risk assessment, using guidance available on official websites.

Those carrying out a fire risk assessment must be satisfied that they are able to

- identify the fire hazards (i.e. potential causes of fire)
- identify people at risk
- evaluate fire safety measures provided and/or required to protect people – for example, escape routes and fire alarm systems
- review the arrangements for management of fire safety – for example, fire procedures, staff training and fire drills, maintenance of fire precautions and so on
- formulate an action plan
- record the significant findings and, if necessary, implement an action plan
- keep the assessment up to date.

Low-risk premises should have simple fire safety procedures in place and completing a fire risk assessment is not complicated. It does, however, need some degree of knowledge about sources of ignition, potential fuel types, likely speed and spread of a fire, human responses to fire, types of evacuation, types of fire precautions needed and fire prevention techniques.

While a competent business owner or manager may be able to complete the fire risk assessment, it would be wise to seek professional assistance from competent individuals or companies. Insurance companies may also be able to assist as it will be in their interests to ensure that the likelihood offire has been adequately assessed and managed so that consequential loss of life and building is kept to a minimum.

Remember that fire risk assessments must record *all* findings and not just those that are significant.

How should the competency of a fire risk assessor be assessed?

It is important and legally required that the person who carries out the fire risk assessment is competent. There are two principal methods by which people can demonstrate their competence

- professional body registration schemes
- certification by a certification body that is UKAS-accredited for the activity.

It is also important that the company for whom the fire risk assessor works has adequate management systems in place, even if the fire risk assessor is self-employed.

Competence of a company to deliver fire risk assessments can be demonstrated by third party certification of the company by a UKAS-accredited certification body.

It is recommended that only fire risk assessment companies, including sole traders, that are third-party certificated to appropriate schemes operated by UKAS-accredited certification bodies are used.

No matter who carries out the fire risk assessment, the duty holder retains the responsibility for ensuring the adequacy of that assessment. If employing a specialist to undertake your fire risk assessment, while you are not expected to be an expert in fire safety, you should make reasonable checks to ensure that they are competent to do the job properly.

There are some simple steps and precautions you can take to help verify the competence and suitability of a prospective fire risk assessor

- be satisfied that the fire risk assessor providing this service is competent to do so
- check that those providing this service have independent registration with, or certification from, a professional or certification body and that they meet the competency criteria established by the Fire Risk Assessment Competency Council. Two such accreditation bodies are

  - Security Systems and Alarms Inspection Board (SSAIB)
  - BAFE Life Safety Fire Risk Assessment (SP205) scheme

- check that they have experience of undertaking fire risk assessments for your kind of business and premises
- check that they have knowledge of the relevant legislation and associated guidance documents
- request references from previous clients in premises of your type; ask these clients if they were satisfied and if any problems were later identified
- check whether they have received any enforcement notices from fire authorities or any written correspondence raising issues with any of their fire risk assessments.

Following a clarification of fire safety legislation in England and Wales, where a building contains two or more sets of domestic premises, the building's structure and external walls are to be assessed in the fire risk assessment. In this scenario the external wall means doors or windows in those walls, and anything attached to the exterior of those walls (including balconies).

Where further technical advice is needed, it must be provided by a competent person as this is critical for ensuring that an appropriate level of safety is achieved. In some cases, the analysis needed may be straightforward, in which case a competent fire safety professional (fire risk assessor) with adequate experience in fire safety and knowledge of external wall systems may be used. In others, it will be more complex and require advice from a qualified engineer with relevant experience in fire safety and fire engineering, including the fire testing of building products and systems, such as a Chartered Engineer registered with the UK Engineering Council.

Approved Code of Practice – A National Framework for Fire Risk Assessor Competency Version 1 – November 2020 – Fire Sector Federation – is a good reference document.

What fire hazards need to be considered?

Consider any significant fire hazards in the room or area under review

- combustible materials – for example, large quantities of paper, combustible fabrics, plastics
- flammable substances – for example, paints, thinners, chemicals, flammable gases, aerosol cans
- ignition sources – for example, naked flames, sparks, portable heaters, smoking materials, hot works equipment.

Do not forget to consider materials that might smoulder and produce quantities of smoke. Also, consider anything that might be able to give off toxic fumes.

Consider the type of insulation involved or used in cavities, roof voids and so on. Combustible material may not always be visible – for example, hidden cables in wall cavities.

What structural features are important to consider when carrying out a fire risk assessment?

Fire, smoke, heat and fumes can travel rapidly through a building if not restricted by fire protection and compartmentation.

Any part of a building which has open areas, open staircases and so on will be more vulnerable to the risk of fire should one start.

Openings in walls, large voids above ceilings and below floors allow a fire to spread rapidly. Large voids also usually contribute extra ventilation thereby adding more oxygen to the fire.

Since the Grenfell Tower fire in 2017, external cladding has become an important element in the assessment of potential fire spread. Many high-rise buildings have been clad externally with materials that are high flammable and which accelerate the spread of fire up the external facades of a building. Fire is then able to leap across cavities onto combustible internal insulation thereby fuelling the rapid spread of the fire in accommodation.

A method of fire prevention is to use fire resistant materials and to design buildings so that fire will not travel from one area to another. Cavity barriers and the fire stopping of voids are critical in containing the spread of fire.

A fire risk assessment must consider the likelihood of fire spread and an assessment must be made of voids and shafts, external cladding and other building materials to establish whether a risk exists.

The risk of fire spread is not restricted to high-rise buildings and therefore all buildings should have an element of assessment of physical fire containment measures. The danger of fire spread is greater in high-rise buildings as the rapid spread of the fire will inevitably restrict and hinder the means of escape of occupants.

Refurbishment works often compromise the original fire stopping in ceiling voids and shafts and the responsible person for fire safety should review all fire safety measures during a refurbishment and construction project, not just for operational fire safety but also for long-term building integrity.

Any opportunity for a fire to spread through the building must be noted on the fire risk assessment.

What are some of the factors to consider when assessing existing control measures for managing fire safety?

Many premises and employers already have some level of fire safety management in place and the Regulatory Reform (Fire Safety) Order 2005 is not intended to add an especially heavy burden on to employers and others as responsible persons.

Existing control measures must be reviewed, and the following are examples of what to look for

- the likely spread of fire
- the likelihood of fire starting
- the number of occupiers of the area
- the use and activity undertaken
- the time available for escape
- the means of escape
- the clarity of the escape
- effectiveness of signage
- how the fire alarm is raised
- can the fire alarm be heard by everyone
- travel distances
- number and widths of exits
- condition of corridors
- storage and obstructions
- inner rooms and dead ends
- type and access to staircases
- openings, voids and so on within the building
- type of fire doors
- use of panic bolts
- unobstructed fire doors
- intumescent strips
- well-fitting fire doors
- propped open fire doors
- type of fire alarm
- location, number and condition of fire extinguishers
- display of fire safety notices
- emergency lighting
- maintenance and testing of fire alarm break glass points
- installation of sprinklers
- location and condition of smoke detectors
- use of heat detectors
- adequate lighting in an evacuation
- training of employees
- practised fire drills
- general housekeeping.

What does an employer need to do regarding fire risk assessments and means of escape for disabled people?

In public places the RRO requires the responsible person, when conducting a fire risk assessment and considering the means of escape from fire, to consider the recommendations of

- the British Standards Institute BS 8300:2018 Design of an accessible and inclusive built environment: Part 1: External environment; Part 2: Buildings. Code of practice.
- Fire safety risk assessment: means of escape for disabled people.

These are not statutory documents but authoritative guidance on the design and management of buildings to enable the safe evacuation of people with disabilities.

The following guidance should be read in conjunction with the British Standard BS 8300:2018 Design of an accessible and inclusive built environment.

The evacuation plan should only be devised by persons familiar with the location and the people involved.

- Disabled people, like everyone else, should always have a safe means of escape in the event of fire.
- The nominated person in charge must, with the assistance of the responsible person, make the best practicable arrangements for ascertaining what areas are used by disabled people, and must, in consultation with them, make adequate arrangements for their evacuation in the event of fire. These arrangements must be tested.
- A personal fire evacuation plan should be drawn up for every disabled person or group of disabled people in the building. Regular building users who are disabled should receive a copy of a personal fire evacuation plan. If the building is one with a large number of visitors, then simple relevant fire evacuation instructions should, so far as possible, be handed to disabled visitors by reception staff.
- So far as reasonably practicable, fire compartmentation in buildings used by disabled people, and any other arrangements, must comply with BS 8300:2018 Design of an accessible and inclusive built environment.
- In most circumstances, if a fire strategy incorporates the use of lifts for evacuation of disabled people, only modern firefighters' lifts or properly designed evacuation lifts conforming to the recommendations of BS 9999, should be used.
- A sufficient number of people should be trained in giving assistance to disabled people so that the necessary number would be present in the event of an emergency.
- Where necessary, arrangements must be made for the presence of the disabled person to be known to those who would give assistance. This could be done with an in-out tally at the entrance or by informing someone, providing the desk or office involved is permanently manned during the day. In some cases, for example ensuring that deaf or blind people are helped, a floor warden system may be more appropriate.
- The placing of restrictions on disabled people requiring them to be accompanied at all times by potential helpers should, where possible, be avoided. In buildings with good fire compartmentation, it will usually be possible for people to work unaccompanied, provided there are adequate numbers of potential helpers elsewhere in the building. However, disabled people who would need assistance to leave in an emergency should not use buildings at times when there are insufficient helpers present to assist evacuation – for example, evenings and weekends. Also, if compartmentation in one area does not reach the standard of BS 8300:2018 Design of an accessible and

inclusive built environment, then it may be necessary to require that a disabled person only uses the area when sufficient numbers are immediately at hand. A disabled person might use a particular floor for normal work and other places in the course of their work.

- Disabled people should not use any part of a building where it would be difficult for them, even with help, to escape in the event of fire. Use of basements by wheelchair users, where there is no basement level exit, is likely to be an example of this. Activities that might take place in such areas should be moved to different areas, so far as reasonably practicable, to avoid excluding disabled people.
- In the case of work above ground floor level by people who use a wheelchair or have difficulty with stairs, arrangements should be based on horizontal movement away from fire through fire-resisting doors to an area of refuge. BS 9999 indicates the layout requirements for this. Procedures could be based on the following principles.

  - When the fire bell rings, the disabled person asks assistance from anyone nearby to help in evacuation. The disabled person and helpers wait, without causing obstruction, in a place near the stairs until other occupants have gone down and the disabled person is then carried or helped downstairs. It may be necessary to provide one or more evac-chairs for this.
  - If insufficient helpers are on hand, the disabled person moves to the main stairwell, or another one if this had been considered by prior agreement with the emergency party to be more convenient, unless there are signs of smoke or fire, in which case the stairwell furthest away from the fire is used, and waits in the stairwell for assistance.
  - The emergency party gathers and if the disabled person is known to be in the building, they go the pre-arranged staircase or, if that is in or very near the fire, to the alternative staircase and carry the disabled person down.

- A fire safety adviser can help in the application of this code to particular circumstances and should be consulted in any case where it appears that building modifications might be required to provide safe means of escape for disabled persons.
- When considering the category of disabled people, responsible persons should include those temporarily disabled through injury or illness.

What are the recommended methods for carrying out a fire risk assessment in high-rise buildings?

In January 2020 the government issued an advice note ‘Advice for building owners of multi-storey, multi-occupied residential buildings’ to help landlords, owners and managers address the need for comprehensive fire risk assessments in high-rise residential buildings. The advice document addressed the concerns with external cladding identified by the Grenfell fire in 2017.

In 2012 the Local Government Association issued a comprehensive guide called ‘Fire safety in purpose-built blocks of flats’, and this document has been the principal guide on how to conduct risk assessments in blocks of flats. The government document published in 2020 refers to the LGA guide and both should be consulted to obtain a comprehensive overview of fire safety and fire risk assessments in residential buildings. Both documents are available free of charge on the government and LGA websites.

There are four types of fire risk assessment for high-rise buildings, depending on the complexity of the building and knowledge of the building materials. Since the Grenfell fire, it is imperative that owners, landlords and property managers consider the risks of fire spread from external wall cladding and insufficient fire doors to individual dwellings. There is also a need to assess any insulation used within the external framework of buildings as a risk is now accepted that fire can spread via insulation and 'jump' from storey to storey.

Type 1 – Common parts only (non-destructive)

A Type 1 fire risk assessment is the basic fire risk assessment required for the purpose of satisfying the RRO.

The inspection of the building is non-destructive. But as well as considering the arrangements for means of escape, the fire risk assessment includes examination of at least a sample of flat entrance doors. It also considers, so far as reasonably practicable, the separating construction between the flats and the common parts without any opening up of construction.

However, in this type of fire risk assessment, entry to flats beyond the area of the flat entrance door is not involved.

Where there are demountable false ceilings in the common parts, it may be appropriate to lift a sample of readily accessible false ceiling tiles. In addition, it will normally be appropriate to open a sample of service risers, provided access is practicable at the time of inspection.

Unless there is reason to expect serious deficiencies in structural fire protection – such as inadequate compartmentation or poor fire stopping – a Type 1 inspection will normally be sufficient for most blocks of purpose-built flats. Where doubt exists in relation to these matters, the action plan of a Type 1 fire risk assessment may recommend that one of the other types of fire risk assessment be carried out or that further investigation be carried out by specialists.

Type 2 – Common parts only (destructive)

The scope and objectives of a Type 2 fire risk assessment are generally similar to those of a Type 1 fire risk assessment, except that there is a degree of destructive inspection carried out on a sampling basis. This will usually necessitate the presence of a contractor for the purpose of opening up construction and making good after the inspection.

In order to check the integrity of separating construction, the areas in which destructive inspection is carried out might sometimes include a sample of flats. However, because of the nature of the work, this can often only be carried out in vacant flats.

A Type 2 fire risk assessment is usually a one-off exercise, which is carried out only if there is good reason to suspect serious structural deficiencies that could lead to spread of fire beyond the flat of fire origin. The age of the block alone is not generally sufficient to warrant a Type 2 inspection. The need for a Type 2 fire risk assessment may sometimes be identified in a Type 1 fire risk assessment but should not simply be recommended as a matter of course.

Type 3 – Common parts and flats (non-destructive)

A Type 3 fire risk assessment includes the work involved in a Type 1 fire risk assessment, but goes beyond the scope of the RRO (though not the scope of the Housing Act). This risk assessment considers the arrangements for means of escape and fire detection (i.e. smoke alarms) within at least a sample of the flats. Within the flats, the inspection is non-destructive, but the fire resistance of doors to rooms is considered.

Measures to prevent fire are not considered unless – for example, in the case of maintenance of the electrical and heating installations – the measures are within the control of, for example, the landlord.

A Type 3 fire risk assessment may sometimes be appropriate for rented flats if there is reason to suspect serious risk to residents in the event of a fire in their flats. (This might be, for example, because of the age of the block or reason to suspect widespread, unauthorised material alterations.) This type of fire risk assessment will not be possible in the case of long leasehold flats, as there is normally no right of access for freeholders.

Type 4 – Common parts and flats (destructive)

A Type 4 fire risk assessment has the same scope of work as a Type 3 fire risk assessment, except that there is a degree of destructive inspection, in both the common parts and the flats carried out on a sampling basis. This will usually necessitate the presence of a contractor for the purpose of opening up construction and making good after the inspection.

However, the nature of the work is such that, often, destructive inspection within flats can only be carried out in those that are vacant. This is the most comprehensive fire risk assessment but will only be appropriate in limited circumstances – such as when a new landlord takes over a block of flats in which the history of works carried out is unknown and there is reason to suspect serious risk to residents from both a fire in their own flats and a fire in neighbours' flats.

(NB: the above four types of risk assessment are reproduced from the LGA guide referenced above.)

What needs to be assessed regarding fire doors to individual dwellings?

A fire door ensures that, should a fire break out, it can be contained in a 'compartment'. This keeps the fire and smoke trapped for a defined period, allowing time for people to get out and make the fire easier to tackle.

It will not fulfil this function if damaged or propped open.

Signs that might indicate a fire door include things like a blue 'Fire door' or 'Keep closed' sign, door closers, intumescent or smoke seals around the edge of the door or the frame.

In blocks of flats, the external door to a flat invariably should be a fire door to protect the common areas from spread of flame and smoke. Other locations will depend on the risk assessment and fire plan of the buildings. Internal doors could well be fire doors, depending on the size of the apartment and distance from the flat entrance door.

All fire doors are fire rated. Some are FD30 (providing 30 min protection), FD60 (60 min protection) or higher. There is usually a certification mark (a label or plug) on top of the door if it is a fire door.

Fire doors should be checked regularly, and the more they're used the more frequently they should be checked.

Anyone can spot a dodgy fire door. But if you have legal responsibility for fire safety, call in a professional.

Create a fire door maintenance checklist and schedule and check all doors in your building.

Only ever replace damaged components with like-for-like. Check the fire certificate. A trained person should be responsible for this maintenance work.

Five point check for fire doors

1. Certification: Check that the fire door is certified. There will usually be a label or plug on the top or side of the door. If the door does not include a label or plug, there is no evidence of certification, report it to the owner of the building.
2. Gaps: The gaps around the top and sides of the door should be no more than 3 mm, plus or minus 1 mm, so no greater than 4 mm and no less than 2 mm. To check if doors meet this standard, use a £1 coin as a guide as this is approximately 3mm wide. In most instances the gap at the bottom of the door can be as much as 8mm. If daylight is seen under the door, the gap is likely to be too big. Where gaps are too big, cold smoke and fire will likely travel through the cracks rendering the fire door non-compliant. Any gaps that exceed this should be reported to the responsible person.
3. Seals: The intumescent seals around the door are designed to expand when they come into contact with heat. This helps to stop the fire from spreading through the cracks and seals the door shut. If the door does not have seals, or if the seals are faulty, report it to the owner of the building or the responsible person.
4. Hinges: Check that all hinges on the door are firmly fixed and that no screws are missing. If a door or the working parts (for example, hinges) are not correctly maintained, it may not be capable of offering the necessary protection in the event of a fire.
5. Closing: A fire door will only offer the necessary protection if it is closed correctly by a door closer or is locked shut as would a store cupboard or riser door. If the door has a door closer fitted, the door can be tested by opening it to about half-way and then letting go. If the door closes fully by itself without catching on the floor or frame, it is fine. If it does not close correctly, report it. A fire door is no use to anyone if it does not close properly or is wedged open.

The fire officer has indicated that my fire risk assessment is not 'suitable and sufficient'. What should I do about it?

If possible, discuss it with the fire officer. Sometimes they do not like the format of a risk assessment rather than the content!

The fire officer may be concerned that you have not considered all aspects of fire safety and that your risk assessment is missing a vital element. Generally, the fire officer should tell you why they are not satisfied with your risk assessment.

If you feel that you have covered everything and that the fire officer is being over-zealous, then ask to discuss the matter with a senior officer.

Be prepared and review your risk assessment together with your competent person or seek professional advice.

A fire risk assessment must address the major fire safety hazards associated with your workplace – it does not have to be perfect.

Case studies

Residential accommodation (Case heard in 2021)

London Fire Brigade inspectors found serious fire safety defects during two visits to a five-storey building in Islington, London.

The building consisted of more than 40 self-contained flats let out to tenants and the inspections revealed deficiencies including

- combustible material near the entrance to the premises
- lack of fire extinguishers and signage
- doors wedged open
- holes within the walls and a non-functioning fire alarm.

The building owner was charged with ten separate offences under the RRO, and the company entered guilty pleas to all charges.

The judge found that the company fell far short of the appropriate standards and there was serious and/or systemic failure within the organisation to address risks.

The Assistant Commissioner for Fire Safety said

> The combined effect of the deficiencies was that if a fire had started, there was a risk of an uncontrolled spread of heat, and smoke and flames affecting the whole premises, coupled with the only means of escape being overcome with smoke.

The building owner was charged with the below offences, entering a guilty plea to all.

Failure to

- take general fire precautions
- have a suitable and sufficient fire risk assessment
- ensure appropriate fire arrangements
- ensure exit routes were clear and unlocked
- ensure appropriate fire safety measures in place and conveyed to employees and residents
- maintain and ensure fire safety equipment
- provide employees with sufficient fire safety information
- provide adequate safety training
- ensure adequate structural compartmentation.

The company was fined over £60 000 with £20 000 costs.

Restaurant owner fails to comply with RRO (2020)

In July 2020, at Derby Crown Court, a director and employer of a restaurant was sentenced in respect of four breaches of the Regulatory Reform (Fire Safety) Order 2005.

Mr *** was sentenced to 18 months imprisonment, reduced to 12 months for an early plea and suspended for two years. Mr *** was also ordered to carry out 240 h unpaid community service and to pay a victim surcharge.

In January 2020, Mr *** entered a guilty plea to all four breaches relating to there being no fire risk assessment, inadequate firefighting equipment, inadequate means of escape and non-compliance with an Enforcement Notice issued in respect of the building.

In the early hours of Sunday 3 December 2017, following a serious fire in a neighbouring property, it was necessary for firefighters to enter the premises known as ***, where they discovered persons sleeping on site in the upper rooms.

As a result of this discovery, a fire safety audit was carried out the following day. The fire safety audit identified that the fire alarm and emergency lighting were defective, general fire precautions were poor, fire extinguishers were out of test date and found to be empty, and it appeared that staff were inadequately trained. In addition, there was no evidence of a fire risk assessment.

These deficiencies resulted in a prohibition notice being served, preventing further use of the property for sleeping until fire safety measures were improved, and an Enforcement Notice requiring other fire safety improvements to the premises in general be made.

Sufficient remedial work was undertaken to allow the fire authority to withdraw the Prohibition Notice, but Mr *** failed to comply with the Enforcement Notice by the required date.

Derbyshire Fire and Rescue Service Group Manager said

> If a fire had occurred in the building where people were sleeping, the absence of working fire detectors and alarms would have allowed the fire to develop, placing occupants at serious risk as smoke and gases from the fire could have killed them while they slept.
>
> Employers, managers and owners of similar establishments should recognise that they have a responsibility to ensure that any sleeping accommodation they provide for their employees, guests, or any other people has a suitable and sufficient fire risk assessment in place, which is reviewed regularly, and facilities that are provided for safety from fire are maintained to a suitable standard.

London Hotel

In March 2015, the former owner of a hotel in Bayswater, London, was fined £200 000 for breaches of the RRO.

When fire safety officers visited the six-storey, 18-room hotel to carry out a routine inspection, they found

- inadequate fire detection systems
- blocked emergency exits
- inadequate emergency lighting
- fire doors missing or tied open with extension cords or string
- the basement room used for sleeping
- no risk assessment had been carried out.

The owner was issued with an enforcement notice. They failed to comply with it. The hotel continued to operate without a working fire detection system. They were fined £200 000 plus £30 000 court costs for seven breaches of the 2005 Order. They were also sentenced to four months' imprisonment, suspended for 18 months.

Fire risk assessor prosecuted (2021)

A fire risk assessor who provided an inadequate assessment for a residential block in Southampton has been fined and given a suspended prison sentence, following a prosecution brought by Hampshire and Isle of Wight Fire and Rescue Service.

At a hearing at Southampton Crown Court, a director of *** Fire Consulting Ltd pleaded guilty to providing a fire risk assessment that was not suitable or sufficient for a fully occupied block of flats.

The director was fined £2750, ordered to pay costs of £19 952 and a given a three-month prison sentence suspended for 18 months. *** Fire Consulting Ltd was fined £20 000 and ordered to pay costs of £19 952 for the same offence.

The fire risk assessor had not lifted ceiling tiles or even opened riser cupboard doors to check for fire safety risks at the three-storey building.

During a visit to Cranleigh House, a Hampshire and Isle of Wight Fire and Rescue Service fire safety inspector also discovered electrical wiring penetrating the compartment walls. Compartmentation is critical in a residential property with a 'stay put' policy and in the apartment block there was no apparent fire stopping added to holes in the walls, meaning that smoke or fire could travel into the communal escape routes to other parts of the building.

His Honour, Judge Burrell QC said

> The job of a fire risk assessor is a highly responsible one. Lives are in their hands and their judgement is crucial. It is not a job to be taken lightly. It is important to hold risk assessors to account and these are serious breaches. I find it odd that there exists no regulatory framework regarding fire risk assessors.

The managing agent for the premises had pleaded guilty at an earlier hearing to failing to comply with an enforcement notice to provide a suitable fire risk assessment. They were fined £10 000 with £6000 costs. The individual manager specifically responsible for the premises was given a conditional discharge and ordered to pay £1800 costs.

The fire safety order places the onus on the responsible person to comply with the legislation. However, in this case, the property management company had put its trust in *** Fire Consulting Ltd to complete the fire risk assessment to a satisfactory standard.

Speaking after the case, the deputy chief fire officer said

> Fire risk assessments underpin the whole process of building fire safety. The assessor failed to inspect and identify fire safety deficiencies within the building and failed to note a compromised alarm and evacuation strategy for the residents.

KEY RESOURCES

Regulatory Reform (Fire Safety) Order 2005
Fire Safety Act 2021
Fire Safety (England) Regulations 2022
Building Safety Act 2022
BS 9999:2017 Fire safety in the design, management and use of buildings. Code of practice
BS 8300:2018 Design of an accessible and inclusive built environment
The fire safety risk assessment guides published by gov.uk:

- Fire safety risk assessment: animal premises and stables
- Fire safety risk assessment: means of escape for disabled people
- Fire safety risk assessment: open-air events and venues
- Fire safety risk assessment: transport premises and facilities

- Fire safety risk assessment: healthcare premises
- Fire safety risk assessment: residential care premises
- Fire safety risk assessment: theatres, cinemas and similar premises
- Fire safety risk assessment: 5-step checklist
- Fire safety risk assessment: factories and warehouses
- Fire safety risk assessment: large places of assembly
- Fire safety risk assessment: small and medium places of assembly
- Fire safety risk assessment: educational premises
- Fire safety risk assessment: sleeping accommodation
- Fire safety risk assessment: offices and shops

Approved Code of Practice – A National Framework for Fire Risk Assessor Competency Version 1 – November 2020 – Fire Sector Federation

PAS 9980:2022 Fire Risk Appraisal of External Wall Construction and Cladding of Existing Blocks of Flats. Code of practice

PAS 79:2012 Fire Risk Assessment. Guidance and recommended methodology

FIRE RISK ASSESSMENT

Name of premises: ______________________________

What particular area of the premises are you reviewing for this Fire Risk Assessment?

What activity, practice, operation and so on are you particularly reviewing for this Fire Risk Assessment?

What ignition sources have you identified?

What sources of fuel have you identified?

Are there any 'extra' sources of oxygen, or will mechanical ventilation increase oxygen levels?

Does anyone do anything that will increase the risk of a fire starting?

If a fire were to start, who would be at risk?

Would anyone be at any extra or special risk, or would any injuries / consequences of the fire be increased?

What precautions are currently in place to reduce the likelihood of a fire occurring, or to deal with it/control it if a fire did start – for example, checks, alarms, emergency procedures and so on.

What other precautions need to be taken, if any. Does anything need to be done to improve existing fire precautions?

How will the information in this Risk Assessment be communicated to staff? Will any staff training take place?

Is there anything else that you think needs to be recorded on this Risk Assessment?

After having identified the hazards and risks of a fire starting within the premises and after considering all the procedures you currently have in place, do you consider the risk to life safety of either staff or customers (including any contractors, visitors and so on) to be:
High ☐ Medium/Normal ☐ Low ☐

If risks to life safety are very high or medium, steps MUST be taken to improve fire safety.

If you implement the other additional measures identified in this Fire Risk Assessment, will risk to life safety of either staff or customers (plus others) be:
High ☐ Medium/Normal ☐ Low ☐

If risks to life safety are possible or very likely, then greater control measures MUST be implemented.

Risk Assessment completed by: ______________________
Date: ______________
Fire Risk Assessment needs a review on: __________________

FIRE RISK ASSESSMENT

Name and address of premises: . . . .

Owner/Employer/Person in control: . . . .

Contact details: . . . .

Date of risk assessment: . . . .

Completed by: . . . .

Use of premises/area under review: . . . .

Identification of fire hazards	High	Medium	Low

Identification and location of those at risk
Evaluation of the risks
Significant findings
Actions taken to reduce/remove risks
Residual Risk Assessment High ☐ Medium/Normal ☐ Low ☐
Review of Risk Assessment: Under what circumstances: How often:

FIRE RISK ASSESSMENT

Name and address of premises:				
Area/room/floor under assessment:				
Name of person completing assessment:			Date of assessment:	
Fire hazard	People at risk	Existing control measures	Proposed action to be taken	Date action taken, by whom: signature

FIRE SAFETY RISK ASSESSMENT

Name of premises:			
Date of assessment:			
Name of person carrying out assessment:			
Area of premises being assessed:			
	Yes	No	Don't know
Identification of fire safety hazards			
Combustible materials, furnishings e.g. carpets, curtains, seat coverings			
Smokers and discarded smoking materials (usually external to the building)			
Electric appliances e.g. portable equipment			
Storage of chemicals, flammable aerosols and so on			
Portable heaters, LPG, flambé lamps			
Cooking equipment, ducts, gas pipes			
Accumulations of waste including cooking oil			
Obstruction of vents or cooling systems			
Building works			
Items stored too near to heat sources			
Obstruction of fire exit routes or fire doors propped open			
Comments **(Please describe anything highlighted 'yes' above)** ■ Furnishings and fittings throughout the premises, including public areas, bars, restaurants, event rooms, bedrooms and offices ■ Electrical appliances including televisions, kettles, irons, hairdryers, glass washers, microwaves, other kitchen appliances and computers ■ Chemical storage for cleaning products and storage of CO_2 ■ Restaurant flambé used within the restaurants ■ Cooking equipment includes gas oven stoves, grills and deep fat fryers in all kitchens and canopy extract systems ■ Accumulation of waste in bar, kitchen, housekeeping and refuse areas including discarded smoking material, paper waste and collection of waste cooking oil			

Who might be at risk from the above hazards? Customers Staff Disabled people Contractors			
Are any people *particularly* at risk if a fire should break out e.g. people working alone in plant rooms, visitors and so on. Describe how and why they will be at risk:			

Perry P
ISBN 978-0-7277-6647-2
https://doi.org/10.1680/fsbqa.66472.077

Chapter 5
Fire safety in building design: key principles

Who is responsible for fire safety when designing new or refurbishment projects?

A designer is responsible for ensuring that the plans and specifications they prepare or the advice they give are sufficient to result in the building work complying with the Building Regulations and all other regulations and laws, if the building work were properly completed in accordance with those plans and specifications or that advice.

The design team should accept collective responsibility for ensuring that fire safety principles are incorporated into any new building design or refurbishment project.

The client should ensure that a competent person has been included on the design team to consider all fire safety matters.

Every design discipline should be responsible for considering fire safety for their design and such information should be coordinated by the principal designer under CDM 2015 regulations.

Complex projects may require the appointment of fire safety engineers and other specialist service providers.

The lead designer, usually the architect, should advise the client and take the lead in ensuring fire safety principles are addressed for the project.

The project team must have competent knowledge of the Building Safety Act 2022, especially if they are designing, developing or refurbishing higher-risk buildings.

What needs to be considered in fire and life safety design?

The topic of designing for fire and life safety is a wide-ranging one and good fire safety design encompasses many disciplines. Expert knowledge is required to ensure that the appropriate fire safety detail is considered, and it is good practice to include a fire engineer on the design team. Fire engineering is the application of scientific and engineering principles, rules (codes) and expert judgement, based on an understanding of the phenomena and effects of fire and of the reaction and behaviour of people to fire to protect people, property and the environment from the destructive effects of fire. The role of the fire engineer is enormously important and is often neglected when the design team on a project is being put together.

The following topics are just a few that must be considered on every project

- the science of fire, smoke generation and spread – for example, spread of smoke, heat accumulation and effect of ventilation
- design principles to minimise the risk of smoke and fire
- means of escape in case of fire
- suitability of fire detection systems
- use and occupancy demands of the building
- material specifications and use of fire-resistant materials
- appropriate access to buildings and facilities for emergency services
- the significance of product specification to fire safety and performance as part of a construction system
- protecting building users from hazards during use and maintenance
- the reaction of people to an outbreak of fire
- the designing of temporary works and the erection of external scaffolding around buildings and the potential for external fire spread and smoke/ventilation impedance during the construction period.

The most effective way to capture all the information required is the production of a comprehensive fire strategy document for the building. When the project is completed, the document should be included in the health and safety file for the building, required under the Construction (Design and Management) Regulations 2015.

The legal requirement for a fire strategy document is set out in the Building Regulations

> where a building is erected or extended, or has undergone a material change of use, Regulation 38 of the Building Regulations requires that a package of fire safety information must be assembled and given to the responsible person of the premises.

A fire strategy is commonly used as a means for collating and providing that information – albeit in many circumstances such information is sadly not provided.

Clients commissioning new and refurbishment projects should stipulate at the project outset their requirements for fire safety information and the format they expect it to be delivered by. It should be stressed that an operations manual of how the alarm system works is not sufficient information to provide a holistic approach to the building fire safety strategy.

When should fire safety be considered on a project?

At concept design stage.

Fire safety must be considered at the very beginning of projects because fire safety engineering requirements may dictate the direction and timetable of the project. Fire safety design principles will need to be considered not only for the new build but also for the impact the new build may have on adjoining buildings.

The Building Safety Act 2022 refers to gateways on a project.

Planning gateway one

Gateway one came into force on 1 August 2021 through the introduction of the Town and Country Planning (Development Management Procedure and Section 62A Applications) (England) (Amendment) Order 2021. This requires fire safety matters to be incorporated into the planning stage for higher-risk buildings. The HSE will be a statutory consultee and submission of fire statements will be required when seeking planning approval.

Although the Building Safety Act requires planning gateway one procedures to apply to higher-risk buildings, the principle of preparing a fire statement is good practice for all building designs.

What is a fire statement?

Fire statements will support the consideration of information on fire safety matters as they relate to land use planning matters – for example, site layout, water supplies for firefighting purposes and access for fire appliances.

It is the intention that the information provided within a fire statement is focused and concise, specific and relevant to the development, and proportionate to the scale, type and complexity of the proposal.

A fire statement will evidence that thinking on fire safety matters, as they relate to planning, has been incorporated into the planning application.

What information must a fire statement contain?

Fire statements must be submitted on a form published by the Secretary of State (or a form to similar effect) and contain the particulars specified or referred to in the form, which includes information about (not exhaustive list)

- the principles, concepts and approach relating to fire safety that have been applied to each building in the development
- the site layout
- emergency vehicle access and water supplies for firefighting purposes
- what, if any, consultation has been undertaken on issues relating to the fire safety of the development and what account has been taken of this
- how any policies relating to fire safety in relevant local development documents have been taken into account.

As many fire safety matters relevant to land use planning impact on the external layout of a site, including the spaces between buildings, fire statements will be required to include information on the entire development as set out on the plan which identifies the land to which the application relates which has to be submitted with the application.

The publication of a standard form for this purpose is intended to ensure consistency in the way in which information is provided, as well as ensuring information contained within a fire statement is focused on fire safety matters as they relate to land use planning.

In what circumstances must a fire statement by submitted with an application?

An application for planning permission for development must be accompanied by a fire statement, unless an exemption applies, when the development involves

- the provision of one or more relevant buildings
- development of an existing relevant building
- or development within the curtilage of a relevant building.

Relevant buildings at planning gateway one contain two or more dwellings or educational accommodation and meet the height condition of 18 m or more in height or seven or more storeys.

What needs to be considered to satisfy the requirement to consider fire safety during the design phase?

Every project will be different and the risk of fire within and around the building will vary depending on what is being built, how the building will be used, occupancy numbers, what materials will be use and many more considerations.

The following list is provided to give readers an overview of the most common elements that need to be considered. It is not an exhaustive list. A fire engineer should be part of the design team and their expert views, opinions and experience should be considered during wide-ranging design discussions. They may have experience of something that designers are unaware of.

Evacuation

In most buildings, the evacuation strategy will be for occupants to evacuate a building immediately on becoming aware of a fire.

In some larger buildings the evacuation strategy allows those occupants most at risk to be evacuated first. This evacuation strategy relies on the building having additional fire safety measures installed in the building and may either be

- phased, where some occupants evacuate parts of the building before others; or
- progressive horizontal evacuation, where occupants leave the compartment of fire origin to adjacent compartments leading to a storey exit.

Designers must have considered the practicalities of evacuation and be able to demonstrate the reasons for the decisions they have taken.

Fire resistance of building elements

In order to prevent the premature collapse or failure of load-bearing structural elements or compartment-separation elements of a building, appropriate levels of fire resistance should be provided.

Occupancy capacity

Is the maximum number of people expected in a space dependent on the area and use of that space – for example, whether the occupants will be seated, standing and so on?

Number of exits

When a room or storey requires two or more escape routes, it is assumed that in the event of a fire one of the escape routes may be compromised by fire. Consequently, the remaining exits will still allow the occupants sufficient time to safely leave the storey without delay.

Inner room

This means a room from which escape is possible only by passing through another room, known as an access room. Occupants within an inner room could become trapped where there is an outbreak of fire in the adjoining access room.

Securing the building

Doors used for means of escape should be kept unlocked at all times when people are in the building. Removable security fastenings, such as shutters, chains, bars, padlocks and so on, should be removed from all doors on exit routes when the building is occupied to ensure the occupants' opportunity for escape is not compromised.

Passive fire safety measures

These depend on activation of fire detection systems – for example, fire doors, dampers, fire shutters, magnetic hold open devices and so on.

Additional fire safety measures

In some buildings it may not always be possible to achieve the minimum standards set in the technical handbooks or guidance. In such circumstances additional compensatory factors may have been used to achieve a satisfactory level of fire safety – for example, additional compartmentation provided, smoke control or pressurisation systems, smoke curtains and so on.

Commissioning certificates

Any certificates and supporting test results should be complete and in a recognised format. They should provide clear and concise information to building owner, including manufacturers' operating instructions for all equipment fitted.

Building management

The management of a building is an integral part of a fire strategy. Developers and builders should note the importance of providing the occupants with information on the use of the equipment and on its maintenance.

Relevant person/duly authorised agent

The relevant person must submit their completion certificate when the building is complete, confirming that the building has been constructed in accordance with the relevant building warrant. A fire safety design summary should be provided with completion certificates relating to the construction of, or conversion to, new non-domestic buildings, including extensions to existing buildings. Where the relevant person does not have the appropriate training, knowledge and expertise to be aware of the hazards and risks involved, then the fire safety design summary should be signed by an authorised agent on behalf of the relevant person.

Case study: Inadequate fire doors and concerns over compartmentation revealed by Perth hotel audit three weeks before fatal fire

The hotel in Perth where three people died in a fire was the subject of a range of fire safety recommendations, following a fire safety audit carried out by Scottish Fire and Rescue just three weeks before the fatal fire.

The audit of the hotel listed 21 improvements that needed to be made following an inspection. The findings and recommendations of the audit included

- defects with an inadequate coverage of the emergency lighting system, with a recommendation for a system upgrade
- various fire doors not providing the correct fire resistance – for example, doors with missing strips and seals
- confirmation that fire resistance ratings of fire doors meet compartmentation requirements
- assessment of the fire resistance of walls and ceilings around the stairwell between the first and second floors where their integrity has been breached, with reinstatement if required
- the requirement of a survey of air transfer grilles in the bar and reception, which may allow passage of fire and smoke to the first floor
- assessment to ensure the correct fire resistance where pipework and cabling passes through floors separating storeys used for sleeping accommodation
- fire drills should be carried out periodically and recorded to ensure all employees understand the emergency fire action plan.

Many of the findings of the fire safety audit related to design issues – for example, fire resistance of walls, doors and so on, transfer of smoke and fumes via air transfer grilles, cabling and pipe ducts not sufficiently fire stopped.

Depending on the age of the building, designers and engineers could be investigated in order to establish whether they were competent to carry out the role or complied with all legal duties in respect of design.

As a project manager, how can I ensure that fire safety matters are considered during the design stage of the project?

Communication and coordination between all design disciplines is the key to a successful project because it will iron out any design clashes, confirm or otherwise assumptions, consider material use, occupancy use and constructability of the building. Designers will confer about their designs and will recognise when adaptations need to be made to accommodate others.

A project manager should initiate regular project design coordination meetings, inviting representatives from the client and all design disciplines.

The CDM regulations require that the principal designer takes the lead on coordinating design risk management, health and safety on the project and ensures that information relevant to the project is shared between those who need to know about it. In addition, the principal designer can be responsible for

- co-operating and sharing information with the Building Safety Regulator
- ensuring compliance with the Building Regulations
- complying with the specific regulatory requirements imposed upon them
- ensuring they and the people they employ are competent to do the work they are undertaking.

The principal designer could take the lead on the design coordination meeting. The architect could also take the lead, with the principal designer assisting where necessary.

The Royal Institute of British Architects has published a fire safety compliance tracker process which can be used to ensure that fire safety is coordinated. It lists various steps and tasks which when completed will record the design development process and compliance with Building Regulations Fire Safety: Approved Document B.

Checklists and record sheets are useful aide memoirs for the design team and allow a documented record of decisions made and the reasons why.

The International Fire Safety Standards have published a set of common principles in relation to fire safety. These are

Prevention principle
Safeguarding against the outbreak of fire and/or limiting its effects.

Detection and communication principle
Investigating and discovering of fire followed by informing occupants and the fire service.

Occupant protection principle
Facilitating occupant avoidance of and escape from the effects of fire.

Containment principle
Limiting of fire and all of its consequences to as small an area as possible.

Extinguishment principle
Suppressing of fire and protecting of the surrounding environment.

(Reference: IFSS – International Fire Safety Standards: Common Principles: Safe Buildings Save Lives.)

For every aspect of a building's design, the project design coordination team must consider each of the five principles

Prevention principle

- how likely is this material to burn
- how quickly will fire spread around the building
- what could cause a fire to start
- what will prevent the spread of flames and smoke
- are there alternatives to the materials/products specified
- what could contribute to the rapid spread of flames and smoke
- how should this be managed during the construction phase?

Detection and communication principle

- if a fire were to start how would the occupants know
- how would they raise the alarm
- could the fire be investigated remotely
- are there mechanisms to identify location of the fire
- is there automatic detection
- how should this be managed during the construction phase?

Occupant protection principle

- how will occupants be protected from the fire
- how will they escape
- is the escape route protected from fire
- how long will it take them to evacuate
- is there a place of safety within the building
- how will people with disabilities fare in a fire
- how will escape routes be kept clear, properly illuminated, clear of smoke
- how should this be managed during the construction phase?

Containment principle

- what compartmentation has been designed
- what is the fire protection rating
- how will services affect compartmentation
- will service ducts etc. facilitate the spread of flames, gases and smoke
- has external cladding been considered
- have higher-risk buildings been clearly identified and consideration given to additional precautions
- how should this be managed during the construction phase?

Extinguishment principle

- what suppression system has been designed

- does it cover the whole building
- is it in modular format
- how does it integrate with the alarm system
- how should this be managed during the construction phase?

When a refurbishment project is being undertaken the project manager and principal and lead designers must always consider the effect offire from the construction works and the new build on existing fire safety arrangements.

Example

A simple example of design coordination to illustrate the process to consider fire safety

An existing building requires new ventilation ducts running between floor levels to the roof. From a fire perspective, the design team want to prevent fire spread from floor to floor, which could be achieved with fire dampers at each floor level or fire rating the entire length of the duct. Space is limited, so the project manger organises several meetings with the architect and mechanical engineer to discuss the problem and potential solutions. The fire engineer has fire rating requirements, the architect has spatial requirements and the mechanical engineer has duct construction and the physical size of fire dampers to consider. After several discussions, the agreed solutions will be added to the plans and specifications. At this stage, a final review of the documents will be made to confirm that everything is as agreed. Reviewing the documents is the last step in the process, not the first or only step.

What records need to be kept to demonstrate compliance with fire safety law during the design process?

It is always a good idea to keep comprehensive records of the design process so that decisions taken can be quickly recalled and evidence produced that a competent approach has been taken by all persons involved in the project.

A formal record of design decisions regarding fire safety could be included on the drawings or kept in a schedule of design risk management attached to the drawings.

Designers must remember that should an investigation occur for any reason into their design, records will be the only evidence that they will have to demonstrate their competency and that they knew what they were doing and took all reasonable diligence to comply with the law.

A comprehensive design risk register could be produced for the project, updated after all relevant meetings by the principal designer and circulated to the client and the design team.

The design risk register could identify whether the hazard is appropriate for the construction phase, occupancy phase, cleaning and maintenance stage or demolition phase of the building.

Residual hazards and their anticipated risks must be recorded and handed over to the client in the health and safety file required by the CDM Regulations 2015.

Designers must also remember that Building Regulations require that a fire strategy document be produced for the project.

The Building Safety Act 2022 requires additional information to be provided to the Building Safety Regulator.

Records help with producing all these documents.

TOP TIPS

The EU Fire Safety Guide sets out seven layers of fire safety in buildings

1. prevention
2. detection
3. early suppression
4. evacuation
5. compartmentation
6. structural safety
7. firefighting

Reference: Modern Building Alliance: EU Fire Safety Guide 2019.

KEY RESOURCES

Building Safety Act 2022
Construction (Design and Management) Regulations 2015
Building Regulations Fire Safety: Approved Document B
IFSS – International Fire Safety Standards: Common Principles: Safe Buildings Save Lives
Grenfell Tower Inquiry Report
RIBA – Fire Safety Compliance Tracker
BS 9999:2017 Fire safety in the design, management and use of buildings. Code of practice
BS 7974:2019 Application of fire safety engineering principles to the design of buildings. Code of practice
Modern Building Alliance: EU Fire Safety Guide 2019

Perry P
ISBN 978-0-7277-6647-2
https://doi.org/10.1680/fsbqa.66472.087

Chapter 6
Fire precautions and fire prevention

What is meant by the term 'fire precautions'?

The term fire precautions refers to any measure that is undertaken to minimise the risk of damage from fire to life or property. It includes the implementation and installation of policies and equipment that are designed to prevent a fire, as well as those designed to suppress a fire should one occur.

Employers have duties to protect their employees and others from the hazards and risks generated by their business, including fire. The measures which they adopt are collectively called 'fire precautions'.

The general fire precautions you may need to take should include the following

- a fire-detection and warning system
- a way of fighting a small fire
- safe routes for people to leave the premises
- suitable fire exit doors
- emergency lighting
- suitable fire-safety signs
- training for your staff or anyone else you may reasonably expect to help in a fire
- a management system to make sure that you maintain your fire safety systems.

What is meant by the term 'fire prevention'?

Exactly what's expected – preventing fires from occurring. Understanding the mechanics of how a fire starts will help individuals know what may cause a fire and take steps to prevent it from happening.

Fire prevention is step one in any fire safety management policy – if fires cannot start, there is no risk.

While all fire precautions are important, are there any that are more important than others?

Yes. The main and most important fire precautions in any building, premises or outdoor arena is the means of escape in case of fire.

Many people die in fires because they have been unable to escape the fire or the smoke and fumes have 'contaminated' the means of escape, preventing them from using it.

The fire authority are especially interested in means of escape when they carry out their statutory inspections.

The other most important aspect of fire precautions is actually 'fire prevention' – why do fires start in the first place? Remember the first 'law' of risk assessment – eliminate the hazard!

What are fire doors?

Doors that are fire resisting and self closing are usually referred to as 'fire doors'.

Fire doors are a vital part of a fire defence system as their purpose is to hold back fire and smoke.

They are normally provided to carry out one of two functions

- to protect the integrity of a structural fire compartment by holding back fire and smoke
- to protect the means of escape for the occupiers of the building for a sufficient period of time for them to escape from the building.

Fire doors can give fire protection for periods ranging from 30 min to 4 h.

Fire doors must be close fitting and constructed of material that withstands heat and flame for the period specified.

Fire doors are normally manufactured with intumescent strips around the door edge. This is, in effect, a smoke seal that expands in heat to fill the gaps between the door and door frame.

Fire doors must generally be kept shut; otherwise, their purpose is not fulfilled.

However, some fire doors can be designed to work with automatic door release systems that allows them to remain in the open position until an alarm or sensor triggers them to shut automatically.

Should all doors be fire doors?

It may be exceptionally good practice to have all doors designed to be fire doors, but it may be cost prohibitive.

Fire doors are needed on all fire escape routes and for rooms that have a high fire rating – for example, plant rooms, electrical cupboards, store rooms and so on.

How do I know if a door is a fire door?

Generally, a fire door is thicker and heavier than a normal door and has an intumescent (brush/fuzzy) strip around its edge, or it is within a door frame that has an intumescent strip.

Often, the doorstops are a minimum of 25 mm.

All fire doors should bear the mandatory sign

Fire door – keep shut
Fire door – keep locked

It is sensible to be able to identify fire doors so that all employees know what and where they are and can assist with keeping them closed.

Is emergency lighting a form of fire protection?

Emergency lighting is not really a form of fire protection, but it is a form of fire precaution.

All escape routes, including external ones, must have sufficient lighting for people to see their way out safely.

Emergency escape lighting may be required if lighting to escape routes is inadequate – that is, they are without natural light, used at night or would be without light if the electrical power were interrupted.

Part of the fire risk assessment will be to check the need for escape lighting to premises. Check all escape routes with the lights off. What can be seen? Are emergency exit routes visible? Is the exit route easy to negotiate and follow? Are there different levels, steps and so on that people could fall down if they can't see them?

Sometimes borrowed lighting or street lighting may be sufficient to illuminate an escape route.

If the fire risk assessment determines that the fire escape routes will be difficult to use without additional light, then some form of emergency lighting will need to be provided.

Emergency lighting will need to be 'proportionate' to the risk and the number of people who will need to evacuate. In small premises with less than five employees and no members of the public, emergency lighting might be a powerful torch kept adjacent to the fire exit door or near the fire point.

In most other premises, emergency lighting should be provided, which automatically comes on when the mains lighting circuit fails. This means that when the building falls into darkness because of a power failure, the emergency lighting circuit takes over and a level of light is generated that is sufficient to enable people to leave in safety.

Also, emergency lighting must be provided if this will help shut down process machinery and so on prior to an evacuation.

Emergency lighting should be sited to cover specific areas

- intersections of corridors
- at each exit door
- near each staircase.

What is required in respect of fire detection equipment?

All workplaces must have arrangements for detecting fire as required by Clause 13 of the RRO. This states that

> where necessary (whether due to the features of the premises, the activity carried on there, any hazard present or any other relevant circumstances) in order to safeguard the safety of relevant persons, the responsible person must ensure that
>
> (*a*) the premises are, to the extent that it is appropriate, equipped with appropriate fire-fighting equipment and with fire detectors and alarms; and
>
> (*b*) any non-automatic fire-fighting equipment so provided is easily accessible, simple to use and indicated by signs.

During working hours, fires are often detected through observation or smell, but this may not always cover areas that are remote or where workers are only sporadically present – for example, plant rooms.

Fire detection systems do not need to be complex, but they do need to be effective. Any fire alarm triggered by fire detection equipment must be audible throughout the premises.

Fire detection systems can include

- smoke detectors
- heat detectors
- flame detectors
- 'beam' detectors.

Fully integrated detection systems are becoming increasingly common, and these are linked to fire alarm systems.

Fire alarms can also be raised by activating a manual 'break glass' call point – the glass is 'broken' and the alarm is activated by pressure on the sensor panel. These systems are usually found in smaller low risk environments.

Control panels are sited in easily accessible positions and indicate where a detector has been activated – that is, in what zone. Heat, smoke or flame may have been detected and the fire alarm activated.

Smoke detectors should be interlinked and should operate on both mains and battery powers.

In smaller workplaces a series of interlinked smoke detectors will probably be all that is required for an effective 'early warning system'.

What is required in respect of fire alarm systems?

All work premises and premises to which the public resort must have a means by which the alarm can be raised in the event of a fire.

Raising the alarm gives people warning that a fire has broken out and encourages them to evacuate the building.

The fire warning system needs to be appropriate for the workplace, but it must be effective and available for everyone to use.

In a smaller workplace where people work closely together a verbal system or manually operated klaxon could be used – for example, shouting 'FIRE! FIRE!'.

Fire warning alarms must be heard in all parts of the building and part of the fire risk assessment must cover the effectiveness of the fire alarm system.

Larger workplaces generally require a more sophisticated alarm system, often an electrically operated break glass call point system.

Manual call points are usually a break glass call point where the glass is 'broken' and the 'button' pushed. This sets off the alarm sounders – often a continuous bell or siren.

A break glass call point will usually set an alarm off throughout the building and indicate evacuation. Occasionally, the fire alarm may be 'zoned' so that alarms go off in one area.

Integrated detection and alarm systems will rely on the identification of smoke or heat in an area and when pre-determined temperatures are reached or smoke is detected, the fire alarm will activate. Automatic fire alarms do not need someone to activate them.

However, every automatic detection and alarm system should be supplemented by a manual alarm call system.

All alarm call points should be on fire exit routes. People need to leave the area first and should raise the alarm on their way out of the building. Call points should also be on staircases and final exit routes.

A good practice guide is that no-one should travel more than 30 m to reach a call point.

The audible fire alarm must be heard in all parts of the building and must be different to any other audible systems in use in the workplace.

Fire alarms should operate between 65dB(A) and 75dB(A). The latter noise level is essential if people will be sleeping on the premises.

It is common for the fire alarm system in places of public access – for example, shops, shopping malls, entertainment complexes – to be incorporated into the public address system so that verbal instructions to evacuate the building are given. However, any public address system used as the fire alarm system must be capable of operating in a power failure so a backup, ancillary power source will be required.

The fire risk assessment must consider what type of alarm is provided within the workplace and whether it is adequate.

The needs of all people must be considered, and fire alarms may need to incorporate flashing lights so as to be seen or understood by everyone – for example, people with hearing impairment.

Visually impaired people need to know that a fire alarm has been activated. This is where a public address system may help.

What is automatic fire suppression?

Automatic fire suppression is a system that can detect, extinguish or contain a fire without having to rely on human intervention. In their simplest form, these systems have a means of detection, actuation and delivery.

Fully automatic fire suppression systems take away the need to depend on a person to discover and tackle a fire. Within seconds of a fire occurring, the detection system detects the radiant heat from the fire, automatically actuates the system and delivers the extinguishing agents straight into the heart of the fire.

Types of extinguishing agents are

- water – for example, sprinklers, water spray, water mist
- gas – for example, carbon dioxide, inert gases such as nitrogen and argon, synthetic chemicals
- foam extinguishing
- powder extinguishing
- special extinguishing – for example, kitchen Ansul system.

The choice of extinguishing system will depend on the location and building circumstances. For example, server rooms and electrical installations are protected with gas extinguishing systems to avoid excessive damage to the plant and equipment when the system is activated.

What is a sprinkler system?

A sprinkler system is a fixed fire extinguishing system installed in many commercial buildings. They are normally used for general protection throughout the building and protect contents, fabric and structure from fire by drenching the fire with water.

Sprinkler systems are becoming increasingly recognised as providing effective life safety for persons in a building and the number of fatalities in buildings that are sprinklered is significantly less than for non-sprinklered buildings.

The British Automatic Sprinkler Association describes a sprinkler system as follows.

> All areas of the building to be protected are covered by a grid of pipes with sprinkler heads fitted into them at regular intervals. Water from a tank via pumps or from the mains water supply (depending on mains pressure) fills the pipes.

Each sprinkler head has a sensor that opens when it reaches a specific temperature, spraying water out and down and onto the fire.

The heat and hot gases from the fire are sufficient to activate the sensors of the sprinklers.

Only sprinklers over the fire that have been activated will open to discharge water. Others remain closed until the temperature in the area rises, thereby triggering them.

The localised triggering of the sensor heads depending on temperature rise minimises the damage to areas not affected by the fire.

Sprinkler heads are spaced generally below the ceiling and designed so that if one operates there is still enough water to be supplied to the other heads.

Fire engineers need to calculate the size, length and volume of water the pipes need to work effectively.

Sprinkler heads can be placed in closed roof voids, ducts, lift shafts and so on and in any areas where a fire could start without being detected.

Sprinkler heads can be isolated by various valves thus allowing for maintenance and repair.

What fire extinguishers will be required?

Clause 13 of the RRO states that

> where necessary (whether due to the features of the premises, the activity carried on there, any hazard present or any other relevant circumstances) in order to safeguard the safety of relevant persons, the responsible person must ensure that
>
> (*a*) the premises are, to the extent that it is appropriate, equipped with appropriate fire-fighting equipment and with fire detectors and alarms; and
>
> (*b*) any non-automatic fire-fighting equipment so provided is easily accessible, simple to use and indicated by signs.

(2) For the purposes of paragraph (1) what is appropriate is to be determined having regard to the dimensions and use of the premises, the equipment contained on the premises, the physical and chemical properties of the substances likely to be present and the maximum number of persons who may be present at any one time.

The fire risk assessment should identify what type of fire extinguishers and how many need to be placed in different parts of the building as the fire risk assessment should have identified which type of fire is more likely to occur in an area.

All workplaces must be provided with suitable means of firefighting equipment for use by people in the premises.

When considering what type of extinguisher to provide, it is necessary to identify the type of material likely to burn.

The British and EN Standards state that a minimum of two Class A fire extinguishers on every floor of a building are needed, unless the premises are very small, in which case one may be acceptable. Nine-litre water extinguishers are the most frequently found types, although general purpose foam-filled extinguishers can be used, and other types may be needed, most commonly carbon dioxide extinguishers for electrical fires.

All premises with electrical equipment (which covers most premises) must have at least 2 kg carbon dioxide extinguishers, 5 kg if there is 415-Volt rated equipment.

The number required depends on the size of the area to be protected and whether the business is classed as low or high risk. As a basic rule of thumb, the floor area in square metres should be divided by 200 and rounded up to calculate how many nine-litre water or equivalent extinguishers are necessary.

If there is a sprinkler or automatic suppression system in place, fewer fire extinguishers may be required.

What is meant by classification of fires?

Fires are classified in accordance with a British and European Standard as follows.

Class A: fires involving solid materials where combustion normally takes place with formation of glowing embers

Class B: fires involving liquids or liquefied solids

Class C: fires involving gases

Class D: fires involving metals

Class F: fires involving cooking oils or fats

Class A and B fires

Class A fires usually involve solid materials, usually of an organic matter such as wood, paper.

They can be dealt with using *water, foam or multi-purpose powder* extinguishers.

Water and foam are the most suitable.

Class B fires involve liquids or liquefied solids such as paints, oils, fats.

These fires can be dealt with using *foam, carbon dioxide or dry powder.*

Carbon dioxide fire extinguishers are best used on fires in electrical equipment. (Note: there is no such fire as an 'electrical' fire.)

Class C fires

Class C fires are those involving gases.

Dry powder fire extinguishers are best used on these fires, but the primary importance is to cut off the gas supply. Consideration needs to be given to the possibility of explosion.

Class D fires

These fires involve metal and need to be dealt with by specialist firefighters. Not only will the metals burn but chemical reactions could take place exacerbating the fire risks.

Class F fires

These fires involve cooking oils and fats and should be dealt with using a smothering medium such as a fire blanket. Some chemical extinguishing systems are designed into cooking canopies – for example, Ansul system. These are specialist systems that should be operated under specialist conditions.

How are the different fire extinguishers identified?

All fire extinguishers produced since 1997 comply with the European Standard BS EN 3 and British Standard 5306, which requires the body or casing of all fire extinguishers to be red.

The body of fire extinguishers can have a coloured band, which represents the contents of the cylinder and the type of fire it is to be used on, of approximately 5% of the external area on the front of the cylinder in the vicinity of the instructions for use.

Another way to reduce confusion as to what fire extinguisher is appropriate to use is to display a wall-mounted sign indicating the type of fire extinguisher and the type of fire to use it on.

Any green fire extinguishers noted in premises should be removed as these contain halon gas and are now illegal in the UK.

Where should fire extinguishers be sited?

Fire extinguishers should normally be sited in conspicuous locations on escape routes, preferably near exit doors.

Fire extinguishers should be grouped together to form a '*fire point*'.

Fire points must be easily identified and could be indicated on the fire plan.

Any room or work area that has been identified as being at risk of a fire should be provided with firefighting equipment – for example, print/photocopy rooms should be provided with a carbon dioxide and water or general purpose fire extinguisher.

If fire extinguishers are placed in areas where they are not clearly visible, they must be adequately signed with a suitable safety sign that complies with the Health and Safety (Safety Signs and Signals) Regulations 1996.

Extinguishers should be securely hung on wall brackets with the carrying handle of larger heavier extinguishers 1 m from the floor. For smaller extinguishers (not heavier than 4 kg), the carrying handle should be mounted 1.5 m from the floor.

Where this is impractical, extinguishers should be located on suitable floor stands. In schools and similar locations, it is preferable to use recessed housing, suitable cabinets or extinguisher covers.

Extinguishers should generally be sited in such a way that it is not necessary to travel more than 30 m from the site of a fire to reach an extinguisher (Class A risk). For flammable liquid, cooking oil/fat or electrical risks the travel distance should not exceed 10 m (Class B and Class F risk).

To avoid confusion, all extinguishers installed in any one building or single occupancy should have the same method of operation and if intended for the same function should be similar in shape, appearance and colour.

Do fire extinguishers have to be wall hung?

Where practicable, fire extinguishers should be placed on wall brackets. This ensures that they are visible, not moved easily and being placed at height makes them easier to lift.

Heavier water fire extinguishers (or powder or foam) should be placed approximately 1 m from the floor, with the handle easily accessible. This will ease the lifting process.

Smaller fire extinguishers, such as carbon dioxide, could be placed at slightly higher positions.

If fire extinguishers cannot be wall mounted, they should be placed in conspicuous positions on a *base plate* on the floor. There are proprietary base plate/stands for fire extinguishers. Again, this should prevent them being moved.

What is the legal requirement for commissioning fire extinguishers?

British Standard 5306-3 (Clause 5.1) states all extinguishers should be subject to commissioning on initial positioning. The commissioning of an extinguisher should be carried out by a competent person. Upon removal from its packaging and transit protection, and immediately prior to placing in its designated place, the extinguisher should undergo the sequence of commissioning service actions considering the safety precautions. Upon completion of the above procedures the extinguisher should then be installed and positioned in accordance with BS 5306-8.

The RRO states that the responsible person must

> where necessary in order to safeguard the safety of relevant persons, must ensure that the premises and any facilities, equipment and devices provided in respect of the premises under this Order or, subject to paragraph (6), under any other enactment, including any enactment repealed or revoked by this Order, are subject to a suitable system of maintenance and are maintained in an efficient state, in efficient working order and in good repair.

When is it necessary to have a hose reel?

The fire risk assessment would identify whether a fire hose is necessary. If in any doubt, consult the fire authority. The legal duty on the responsible person is to provide appropriate firefighting equipment for the risks identified in the premises.

A hose reel is a length of tubing with a shut off valve connected to a permanent pressurised water supply.

Hose reels used to be common in high-rise offices, hotels and halls of residence, but increasingly they have been removed in favour of modern, pressurised fire extinguishers, as these are deemed to be more appropriate to tackle localised fires. There is also a potential risk that little-used hoses could be colonised by Legionella bacteria, thus creating a risk of Legionnaires' disease when used.

Fire hoses can sometimes lull people into a false sense of security about fighting a fire and they may be tempted to take risks. Fire hoses should only be used by trained competent people.

Often the fire brigade will use the fire hose as a quicker method to fight the fire than drag their own hoses up several flights of stairs.

If hose reels are provided in any building the responsible person must ensure that the equipment is properly maintained.

When is it necessary to have a fire blanket?

Once again, the need to provide fire blankets will be determined in the Fire Risk Assessment as it comes under the provision of firefighting equipment in the RRO.

Fire blankets are used to smother flames, usually involving cooking oils and grills, chip pans and fryers.

A fire blanket should always be provided in a kitchen. It must be easily accessible and should not be placed behind the equipment it is intended to be used on as the flames may prevent it being accessed.

Fire blankets can also be used in industrial process areas where fires need to be smothered and where molten metals are involved.

Fire blankets are classified as light-duty or heavy-duty.

Light-duty fire blankets are usually suitable for dealing with small fires in containers – for example, chip pans.

Fire blankets can also be used to wrap around burning clothing when flames need to be smothered quickly.

Fire blankets need to be regularly checked and can only be used once.

What is passive fire protection?

Passive fire protection (PFP) is a form of fire safety provision that remains dormant or inert during normal conditions but becomes active in a fire situation.

It is an integral component of structural fire protection in a building, which is designed to contain fires or slow their spread. The purpose of PFP is to contain the spread of fire for sufficient time to permit

- the safe evacuation of all occupants of the premises
- the arrival of the fire brigade.

The person responsible for fire safety also has a duty of care towards any members of the emergency services – for example, firefighters – who may have to enter the premises during the course of a fire; in slowing the spread offlames, smoke and hot gases, PFP also serves to ensure the building remains as safe as possible for entry in this situation.

PFP provision is required in all buildings, whether domestic or non-domestic, with the purpose of containing, compartmentalising and retarding the spread of fire.

In respect of internal fire spread (within a structure) and the relevance of PFP, the Building Regulations stipulate the following (Requirement B3).

> Where reasonably necessary to inhibit the spread of fire within the building, measures shall be taken, to an extent appropriate to the size and intended use of the building, comprising either or both of the following –
>
> (*a*) sub-division of the building with fire-resisting construction.
>
> (*b*) installation of suitable automatic fire suppression systems.

> The building shall be designed and constructed so that the unseen spread of fire and smoke within concealed spaces in its structure and fabric is inhibited.

In respect of the protection of openings and fire-stopping in order to inhibit the spread of fire, Section 9 of the Regulations states

> every joint, imperfect fit and opening for services should be sealed. Fire-stopping delays the spread of fire and, generally, the spread of smoke as well.

What are intumescent seals?

Often, the most hazardous aspect of a fire is the generation and transfer of smoke and hot gases around the building.

An intumescent substance is one that swells because of heat exposure, thus increasing in volume and decreasing in density. Intumescent materials are predominately used in passive fire protection, although a product of any material that expands under the single introduction of heat may be called intumescent.

Intumescent strips are fitted to fire doors and, when exposed to extreme heat, they expand to close any gaps or edges that might facilitate the spread of smoke and fire. The spread of smoke in a fire can cause more risk to escaping occupiers, especially if the smoke penetrates through to escape routes.

Intumescent air transfer grilles, which are typically 30 or 60 min fire rated, allow air to circulate freely around a building under normal conditions, but the intumescent material swells and creates a barrier to restrict the passage of hot gases in a fire situation. They are suitable for use with both fire rated doors and compartment walls.

Intumescent pipe wraps and collars are designed for use on plastic pipes that pass through masonry floors and walls; the intumescent material expands inwards in a fire situation to fill the void of the collapsing pipe.

It is common for the fire stopping integrity of building structures to be breached during refurbishment works. Responsible persons are advised to check all refurbishment plans to ensure that the design team and contractors are conversant with the need for fire stopping and that they specify the correct procedures for maintaining the integrity of the building.

What is lightning protection?

Lightning protection systems are used to prevent or lessen lightning strike damage to buildings. They protect the internal electrical components of a building, helping to prevent fires or electrocution. Lightning protection comes in the form of a lightning conductor, usually a metal rod, mounted on a building to protect it from lightning strikes. The system will intercept a strike so, if lightning hits the building, the lightning rod will be hit first, causing the strike to be conducted through a wire and passed through to the ground.

Lightening protection will help reduce the likelihood of a fire breaking out during an electrical storm.

What is fire stopping?

Fire stopping, also known as compartmentation, is a fundamental part of passive fireproofing. It refers to the process of filling openings and joints between walls and floors with fire-resistant material, inhibiting the spread of fire between compartments within a building. It will usually have a specific fire rating – for example, 30 or 60 min.

Fire stopping to services installations – for example, trunking, duct work, cable trays, cabling, ventilation grilles and so on – is vital to minimise the spread of fire through a building.

The importance of fire stopping

Many studies had been carried out on the effectiveness of fire stopping and in one of the studies presented to officials at the building regulations division at the then Department for Communities and Local Government the researchers said

> barriers intended to stop fires spreading were often found to be missing or incomplete or incorrectly positioned.

This failure emerged later from investigations into the refurbishment of Grenfell Tower as identified in the public inquiry into the disaster.

Another study warned that people had died because of fire or smoke spread in concealed spaces and

> the potential risks and the potential losses remain high.

Inquiry investigations have also found that cavities and vertical channels in Grenfell Tower's new cladding system were probable contributors to the rapid vertical fire spread.

Fire stopping restricts the spread for fire and smoke. It contains the fire in one area.

Case studies: Lack of fire stopping

Block of flats

A fire in a four-storey purpose-built block of flats was discovered at around 00:38. The building comprised four flats on each floor and was constructed in the 1990s from timber frame with predominantly brick cladding.

The fire started in a cavity wall on the ground floor, spreading horizontally and then vertically upwards to the second and third floors and into the roof space where it burned through part of the roof.

Cavity barriers were in place horizontally between floors; however, the oriel window detail at the front of the property did not have any cavity barriers either vertically or horizontally.

These missing cavity barriers created a route for fire spread up the building. There were no reported injuries, but 33 residents were evacuated as a precaution.

The cause of the fire is understood to have been due to an electrical cable arcing on a plasterboard fixing in the wall cavity.

Block of flats

A fire was discovered in a basement car park of a block of flats with four accommodation storeys above at around 04:00.

The cause of the fire was undetermined but fully involved two cars.

Smoke and flames from the fire spread to the upper floors of the building due to poor fire stopping of penetrations in the concrete slab between the car park and the upper floors.

The fire burned through the floor above by way of a poorly fire-stopped soil pipe, into a bedroom of one of the flats where the bed caught fire with the resident asleep in it.

The inadequacy of the fire stopping was attributed to poor workmanship as some fire stopping around penetrations was ineffective.

Three people suffered smoke inhalation and one sustained a broken ankle. Ten other people were rescued by the fire and rescue service.

KEY RESOURCES

Regulatory Reform (Fire Safety) Order 2005
Building Regulations 2010 Fire Safety: Approved Document B (as amended 2022)
BS 5306:2020 Fire extinguishing installations and equipment on premises
BS EN 54-1:2021 Fire detection and fire alarm systems: various parts within the series
BS 5839-6:2019 Fire detection and fire alarm systems for buildings, 4th edition

Perry P
ISBN 978-0-7277-6647-2
https://doi.org/10.1680/fsbqa.66472.103

Chapter 7
Means of escape in commercial and retail premises

What is meant by the term 'means of escape'?

The term means of escape in fire safety refers to the means that are provided in a building for occupants to leave the building in an emergency, quickly and safely without unnecessary assistance.

Means of escape should be provided from any point on a floor to an exit from the floor. The general principle is that any person confronted by a fire within a building can turn away from it and escape safely.

Means of escape are usually alternative exits from a building that have been designed to be accessed quickly from an occupied area and that have been protected in some way from being affected by fire or smoke.

Means of escape should lead to a place of safety, should be easy to identify and safe to use.

Means of escape should, where possible, allow escape from a building in more than one direction and in a direction other than the main entrance.

Are there any exit routes that are not suitable means of escape?

Any escape route that could be hazardous to use would be unsuitable. In particular, the following are considered unsuitable as means of escape

- lifts
- portable ladders
- long spiral staircases
- escalators
- 'throw out' devices or escape ropes and so on.

Why is it not acceptable to use lifts as a means of escape?

Lifts are unsuitable as a means of escape because

- the power may be disconnected because of the fire and occupants of the lift car could be trapped in the lift shaft in a burning building

- lifts could stop and disgorge occupants onto the floor on which the fire is raging. Lift occupants will not know which floor has the fire and could walk into it, becoming trapped.

There are some designs of lifts that can be used in a fire, and these are known as 'firefighter' lifts. They are designed to run on separate power systems and are installed for the use of firefighters so that they can get to the scene of a fire in a tall building quickly.

What is a 'place of ultimate safety'?

A place of safety is a place in which a person is no longer at risk from a fire.

A place of safety must be in the open air and must be a place beyond the building.

Means of escape take people to a place of safety – that is, a location away from the building where they will not be at risk from heat, fumes, smoke, collapsing structures and so on.

It is not adequate for occupants to discharge from the building into an adjoining alleyway, small yard, inner courtyard, rear garden and so on from which there is no escape and where occupants would still be at risk.

Is a place of safety the same as an assembly point?

Not necessarily so. Means of escape lead people *to* a place of safety – that is, an area in the open air beyond the building.

An assembly point is a location at which people are requested to congregate so that a roll call can be made to account for all occupants, employees and so on.

What is a place of 'relative safety'?

Where there is considerable distance between an exit route and the final, external place of safety, it is necessary to provide an intermediate place of safety – known as the place of relative safety. This will be a location where people could rest for a few seconds to gather their bearings, catch their breath or prepare themselves for the final descent to the final exit.

Multi-storey buildings have long distances from the upper floors to the ground exit routes and places of relative safety are provided on protected staircases – that is, it is relatively safe to wait in the staircase pending the next stage of the exit.

What is a secure waiting area?

Secure waiting area means a space in the adjacent fire compartment that is connected to an escape route where persons with reduced mobility or sense of direction can await further evacuation. A secure waiting area may also be a part of the escape route if the secure waiting area is located adjacent to the areas served by the escape route.

The secure waiting area should be able to accommodate people with reduced mobility or sense of direction.

Secure waiting areas should be situated on the same floor as the space it serves.

There should be a possibility for two-way communication from the secure waiting area.

The communication system function should be supported in the event of a power failure and have protection against power outages as a result of the fire.

What are the key principles involved in designing means of escape?

The first fundamental principle of a means of escape is that people should be able to turn their back to a fire and leave the building to a place of safety.

Buildings should preferably have two means of exit, but smaller buildings may not.

Dead ends should be avoided, as people may get to the end of the corridor and have no other way to go. This could mean that they would have to turn back towards the fire to exit the building.

Means of escape should be wide enough to take the occupants of the area without crushing and overcrowding.

Exits from areas of the building should be separate and independent and not too close together that they really do not present an alternative.

Exit routes from upper floors will obviously not lead, initially, to an outside place of safety. In this instance an exit route must lead to a place of 'relative safety'.

What is known as the process of escape?

Having considered the factors that will influence escape and having seen how these can be related to the risk profile and/or occupancy levels of a specific building, it is important to look at the stages in the process of escape and the maximum distances people can be expected to travel.

Escape is generally considered in four distinct 'stages' as follows

- Stage 1: escape from the room or area of fire origin
- Stage 2: escape from the compartment of origin via the circulation route to a protected stairway or an adjoining compartment offering refuge
- Stage 3: escape from the floor of origin to the ground level
- Stage 4: escape at ground level away from the building.

It is important that each floor plan of a building indicates the shortest route(s) to a place of comparative or ultimate safety should an emergency evacuation be triggered – for example, by the sounding of the fire alarm. The width of final exit doors and the escape routes leading to them will dictate the maximum number of people who can safely occupy that floor or a specific area within it under normal conditions of operation.

What is meant by the term 'travel distance'?

Travel distance can be defined as the distance to be travelled from any part of the building to the nearest

- final exit route and place of safety in the open air, away from the building
- door to a protected staircase or other place of relative safety
- door to an escape route located externally.

What is escape time?

In the event of a fire, it is important to evacuate people as quickly as possible from the premises. Escape routes in a building should be designed so that people can escape quickly enough to ensure they are not placed in any danger from fire. The time available will depend on several factors, including how quickly the fire is detected and the alarm raised, the number of escape routes available, the nature of the occupants and the speed of fire growth.

Often, when fire evacuation drills are undertaken, a key performance indicator is how long it took to evacuate the building. Time is of the essence in a developing fire and speed of evacuation is critical to ensuring safety.

What are the legal requirements regarding distance of travel in commercial, retail and hospitality buildings?

Distances of travel are laid down in the Building Regulations 2010 Fire Safety: Approved Document B, in the British Standards applicable to fire safety, namely BS 9999:2017 and in the formal guidance documents issued by the government on fire safety.

The distance of travel in a building depends on the design and use of the building and the assessment of the fire risk.

Single storey buildings

If two exit routes are provided this would be acceptable provided the distance to either exit was not more than

- 45 m in offices and factories
- 30 m in shops and other premises

Exits must not be so sited that they are too close together, making them both inaccessible in the event of a fire.

In order to judge the effectiveness of exit routes it is recommended to measure the angle formed between the two doors.

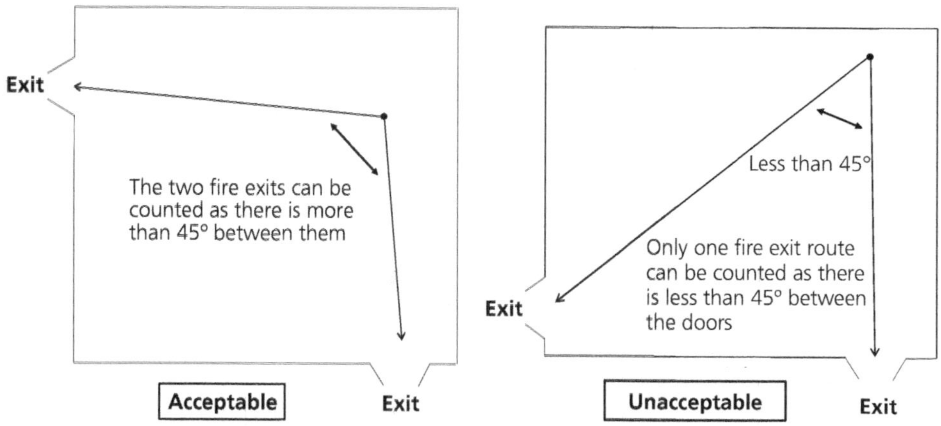

If the two doors form an angle of access between each other greater than 45° they will be deemed to be acceptable as alternative exits. If they are less than 45° they are too close together to give an effective alternative.

Where a floor is subdivided into offices or other rooms, corridors and so on, there is a maximum travel distance from the room to a ‘room exit’ of no more than 12 m.

In addition, the overall distance of travel to the final exit must not exceed 45 m in offices and factories and 30 m in shops.

If only *one* exit is provided to a single storey building the distance of travel to a final exit will be

- a maximum of 18 m in offices and shops
- a maximum of 25 m in factories.

Multi-storeyed premises

Distance of travel criteria that relate to single storey buildings should be followed for multi-storeyed buildings.

Stairways/staircases serving upper and lower floors must be ‘protected’ – that is, have some form of fire-resistant construction.

The distance of travel to a protected staircase must be

- a maximum of 18 m in offices and shops
- a maximum of 25 m in factories.

The distance to be measured is from the furthest point of the floor to the entrance to the protected fire staircase.

If the staircase is *not* protected, then the distance of travel from the furthest point of the floor to the *final* exit route on the ground floor must be

- a maximum of 18 m in offices and shops
- a maximum of 25 m in factories.

If a building is provided with more than one staircase, then the distance of travel to each staircase must be

- in unprotected staircases
  - 45 m in offices and factories
  - 30 m in shops

provided those distances lead to the *final exits*

- in protected staircases
  - 45 m in offices and factories
  - 30 m in shops

provided those distances lead to the exit route leading into the protected staircase.

Does the distance of travel calculations vary from building to building?

The distance of travel calculations will be affected by the risk rating of the premises in respect of fire.

If the danger to life safety is considered to be high, then there will be a necessity to evacuate people from the building very quickly. Therefore, the shorter the distance of travel to a place of safety, or place of relative safety, the better the fire precautions for the building.

Travel distances may need to be reduced to

- 12 m in high-risk shops to a storey exit in one direction
- 25 m in high-risk shops to a storey exit in more than one direction
- 6 m from within a room to the room exit in high-risk shops or offices.

Travel distances should be considered when preparing the premises' fire risk assessment. Travel distances may be longer than those recommended in guidance documents if site specific criteria reduce the risk of fire or where persons – for example, employees – are very familiar with the premises and exit routes are clear, unobstructed and easy to navigate.

How is the distance of travel measured?

Distance of travel should be measured as being the actual distance to be travelled between any point in the building and the nearest storey exit.

The distance should not be 'as the crow flies' but should measure the actual route to be taken and must include detours around fixtures and fittings and so on.

What is an access room?

Where the only means of escape from a room is through another room, the outer room is called an access room.

What is an inner room?

An inner room is a room that is entered via an outer room or access room.

An inner room must be under the control of the access room occupier.

Inner room and access room

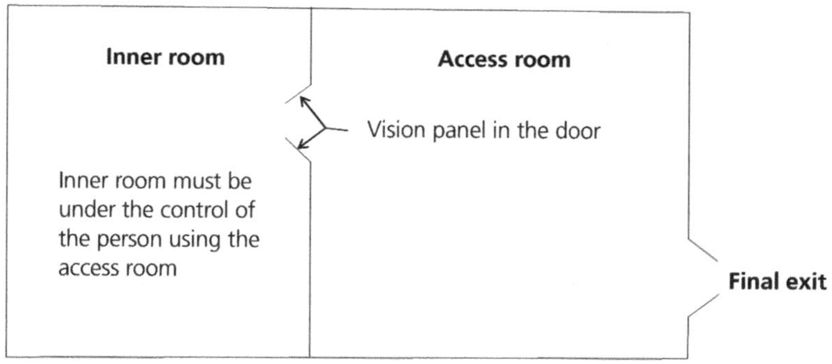

Why are these rooms important for fire safety?

Account must be made of a fire starting in an access room and how someone in the inner room would be able to evacuate. A fire in the access room may restrict those in the inner room from leaving the room.

Evacuation will be easier and quicker if the occupants of the inner room have clear visibility into the access room so that they can see either the beginnings of a fire or be made aware of the dangers by the people occupying the access room.

An inner room therefore must have suitable vision panels in any doors or partitions.

Occupants of the inner room must be able to detect the early stages of a fire.

An inner room should not be located off an access room that has a high fire risk rating – for example, stores hazardous/flammable chemicals, kitchen and so on.

The distance of travel from an inner room to the exit from the access room should not exceed 12 m.

Under certain circumstances an inner room may not be provided with vision panels – for example, sanitary accommodation, changing facilities. Generally, these rooms would not be places of work and would not have occupancy of long duration.

It is *not* acceptable to have an inner room entered via an access room which is itself entered via an access room.

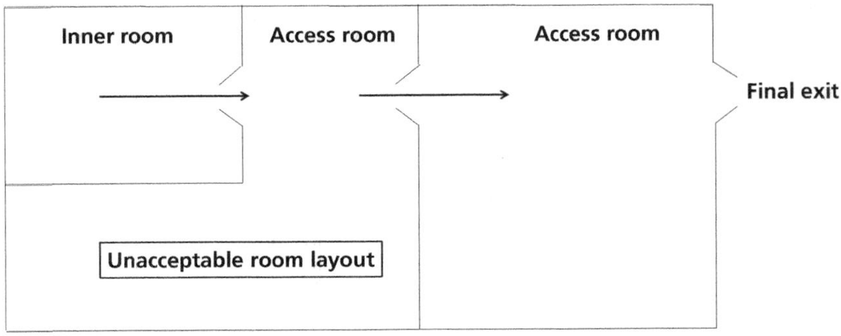

Do means of escape corridors have to be sub-divided to prevent the spread of fire?

Corridors exceeding 30 m in length in shops and 45 m in lengths in offices and factories should be sub-divided by fire doors to prevent the free travel of smoke, heat and toxic fumes.

The sub-division of corridors should be such that no individual length of corridor is common to more than one storey exit.

What is meant by the terms 'protected staircase' and 'protected lobby'?

A staircase is protected if it is enclosed by fire-resisting construction materials.

Fire-resisting materials are those that will withstand the onslaught of fire for a minimum of 30 min. Greater fire resistance is required in certain locations – for example, public buildings.

A staircase is only protected when it is protected from fire in adjoining accommodation.

A protected staircase must discharge to a final exit or lead onto a protected route that then leads to the final exit.

A protected lobby is a fire resisting enclosure providing access to a protected stairway via two sets of fire-resisting self-closing doors and into which no rooms open, other than toilets or lifts.

What is a protected route?

Protected routes are escape routes that are physically separated from other parts of the building by fire-resisting construction to minimise the risk of the escape route becoming unusable due to fire and smoke spread within the building.

In practice fire resistance is usually to a minimum of 30 min fire (in some cases it can be much more) and 'construction' is usually walls, ceilings, glazing, doors and so on.

Once in a protected route, it should lead to a final exit to a place of safety. You should not usually have to leave a protected route to then go through an unprotected one to reach an exit.

When do the principles of travel distances, protected routes, inner rooms and so on need to be applied to existing buildings?

Generally, when an occupier, owner or employer is considering internal alterations and refurbishment works, they must take into consideration the fire safety requirements of the alterations.

Often, erecting internal partitions creates inner rooms leading off access rooms, or partitions will obstruct means of escape routes by increasing the travel distances.

It would be sensible to consider the existing means of escape before carrying out works and to plan the remedial works in relation to the principles of good fire safety.

The fire risk assessment must be reviewed whenever alterations are proposed and any material alterations to the building must be incorporated into a revised fire risk assessment.

What are fire doors?

Door that are fire-resisting and self-closing are usually referred to as 'fire doors'.

Fire doors are a vital part of a fire defence system as their purpose is to hold back fire and smoke.

They are normally provided to carry out one of two functions

- to protect the integrity of a structural fire compartment by holding back fire and smoke
- to protect the means of escape for the occupiers of the building for a sufficient period for them to escape from the building.

Fire doors can give fire protection for periods ranging from 30 min to 4 h.

Fire doors must be close fitting and constructed of material that withstands heat and flame for the period specified.

Fire doors are normally manufactured with intumescent strips around the door edge. This is a smoke seal that expands in heat to fill the gaps between the door and door frame.

Fire doors must generally be kept shut otherwise their purpose is not fulfilled.

However, some fire doors can be designed to work with automatic door release systems that allow them to remain in the open position until an alarm or sensor triggers them to shut automatically.

Should all doors be fire doors?

It may be exceptionally good practice to have all doors designed to be fire doors, but it may be cost prohibitive.

Fire doors are needed on all fire escape routes, to rooms that have a high fire rating – for example, plant rooms, electrical cupboards, storerooms and so on.

How do I know if a door is a fire door?

Generally, a fire door is thicker and heavier than a normal door and has an intumescent (brush/fuzzy) strip around its edge, or it is within a door frame that has an intumescent strip.

Often, the doorstops are a minimum of 25 mm.

All fire doors should bear the mandatory sign

Fire door – keep shut
Fire door – keep locked

It is sensible to be able to identify fire doors so that all employees know what and where they are and can assist with keeping them closed.

Is there a minimum width for fire exit doors?

The width of fire doors is determined by the number of occupants of a room or building.

The minimum width for an emergency escape exit should ideally be 1050 mm but not less than 750 mm in any case and where wheelchair users are likely to use it not less than 900 mm.

The premises' fire risk level will be obtained from your fire risk assessment as shown in the table below. Fire risk will depend on use of the building, fire risk, occupancy numbers.

Minimum width of escape route or exit	Building fire risk and persons
750 mm – 1050 mm	Low risk = 80 persons Medium risk = 100 persons High risk = 120 persons
1050 mm	Low risk = 160 persons Medium risk = 200 persons High risk = 240 persons
75 mm extra over 1050 mm for every 15 persons regardless of risk level	

When calculating the exit capacity, one must remember to discount the largest exit or exits, if they are close together (see example below), as they may be unavailable due to the fire. Then add together the number of persons who can safely use the remaining available exits.

Example

Exit A (1200 mm wide) = 200 + 30 (150 mm over 1050 mm) = 230 persons, Exit B (900 mm wide) = 100 persons and Exit C (850 mm wide) = 100 persons. Discount the largest Exit (A) then total exit capacity is Exit B + Exit C = 200 persons.

How many fire exits do I need?

There is no specific regulation defined by the government on how many fire exits you must have in your premises; however, there are guidelines in the Building Regulations 2010 Fire Safety: Approved Document B to follow to comply with fire safety regulations.

The government recommends that more than one fire exit is provided from your premises whenever it is possible. These exits should be completely independent from each other and have separate escape routes if possible.

Three factors influence the number of fire exit doors needed in a premises

- the number of occupants in the room
- the number of tiers or storeys in a building
- the limits on travel distance to the nearest exit.

The government guideline states that for up to 60 persons one fire exit is needed, for up to 600 persons two fire exits are needed and in cases of more than 600 people there must be at least three fire exits. It is not always possible to know the number of people in the room, so in this instance the capacity of the room is calculated based on the floor space factor found in Appendix C of Building Regulations 2010 Fire Safety: Approved Document B.

The distance people need to travel to evacuate the premises must be as short as possible. However, if a building has more than one exit, there should be no point that is more than 60 m away from the nearest escape route.

The responsible person must make sure that the premises reach the required standards and that the occupants of the building are provided with adequate fire safety guidance.

The RRO applies to almost all buildings, places and structures, including shared areas in houses in multiple occupation (HMOs), blocks of flats and maisonettes; however, individual private homes are not covered in the RRO.

What is required regarding the maintenance and testing of fire escape doors?

Article 17 of the RRO makes it a legal requirement to ensure that fire-resisting doors and escape doors are correctly installed and adequately maintained for them to be fit for purpose.

Maintenance and testing of routes in case of an emergency are not just limited to maintaining fire escape doors but also staircases, corridors, escape routes, fire detection alarm systems, firefighting equipment, notices and emergency lighting. Regular checks, periodic servicing and maintenance must be carried out on all fire safety fixtures and fittings.

An employer or nominated employee can be designated to carry out checks and routine maintenance of emergency routes and equipment. However, it is recommended that periodic servicing of firefighting equipment and installed systems, such as fire alarms and lighting, are done by a competent person to ensure reliability, safe operation and that any necessary repairs are carried out.

Every premises will have its own requirements depending on the size and layout of the building and the number of people inside.

- *Daily*: responsible persons or wardens should check that all escape routes are kept clear, that fire doors open easily and that there are no fire hazards such as visibly faulty electrical equipment or flammable materials left close to a heat source.
- *Weekly*: test fire alarms to ensure the mechanisms are working correctly.
- *Six monthly*: UK safety standard BS 5839 recommends that fire alarm systems are inspected by a 'competent person' at least every six months.
- *Annually*: all fire extinguishers should receive a basic service every 12 months, to pick up on any problems and ultimately to ensure that the extinguisher has a long and reliable life. Fire suppression systems also need regular servicing, timings of which will vary depending on specific products and the environments in which they're stored.
- *Five yearly*: most fire extinguishers should have an extended service after five years. This will involve completely discharging the extinguisher, checking for internal corrosion, refilling and re-pressurising.

Fire Safety (England) Regulations 2022 introduced new legal duties for the responsible person to carry out quarterly checks on flat entrance doors leading onto common parts of multi-occupied residential buildings over 11 m. This requirement has been introduced as a result of the Grenfell Tower fire and recommendations made in the public inquiry report.

Responsible persons must make a record of their attempts to check flat entrance doors.

Can I leave a fire exit door open?

If it is a hot day or some airflow is needed, it is completely fine to have final fire exit doors standing open. The only time this shouldn't happen is if the fire exit door is also doubling up as a fire-resistant door to stop the spread of fire. Fire rated doors – that is, those with either 30 or 60 min fire resistance – should not be left open.

Can fire exit doors be locked?

Internal fire exit doors

There are many instances where emergency escape doors need to be locked. For example, hotel bedroom doors must have the ability to be locked for the privacy and security of the guest and are operated externally using a key, but can be readily opened from the inside for easy evacuation in an emergency. Building Regulations state

> where a door on an escape route has to be secured against entry when the building or part of the building is occupied, it should be fitted with a lock or fastening that is readily operated, without a key, from the side approached by people making their escape.

Similarly, if the door is operated by a code, card or biometric data, for example, it should be possible to override the lock by the side of the door used by those making their escape. This stipulation is further reflected in the RRO by stating

> emergency doors must not be locked or fastened that they cannot be easily opened by any person who may be required to use them in an emergency.

External fire exit doors

Similarly, it is also possible to fit an outside access device (OAD) to an external fire escape door. Fitting an OAD enables the door to be made accessible externally without fitting a lock internally that would inhibit the use of the door in an emergency.

In non-residential buildings some fire exit doors may be fitted with a lock that is only used when the building is empty; however, the onus is placed on the management procedures to ensure the safe use of locks in this situation.

A breach of fire safety legislation has far-reaching implications. In addition to any potential fines, there is also a loss of business confidence and a risk of causing injury or even death through failure to safeguard employees or tenants. In the unfortunate instance that a fire occurs on your premises, you must show that as far as reasonably practicable you put measures in place to negate the impact of such an incident. Ignorance is no defence, and no company is exempt from prosecution.

Is emergency lighting required on escape routes?

All escape routes, including external ones, must have sufficient lighting for people to see their way out safely.

Emergency escape lighting may be required if lighting to escape routes is inadequate – that is, they are without natural light or daylight, used at night or would be without light if the electrical power were interrupted.

Part of the fire risk assessment will be to check the need for escape lighting to premises. Check all escape routes with the lights off. What can be seen? Are emergency exit routes visible? Is the exit route easy to negotiate and follow? Are there different levels, steps and so on down which people could fall if they can't see them?

Sometimes borrowed lighting or street lighting may be sufficient to illuminate to escape route.

If the fire risk assessment determines that the fire escape routes will be difficult to use without additional light them some form of emergency lighting will need to be provided.

Emergency lighting will need to be 'proportionate' to the risk and the number of people who will need to evacuate. In small premises with less than five employees and no members of the public, emergency lighting might be a powerful torch kept adjacent to the fire exit door or near the fire point.

In most other premises, emergency lighting should be provided that automatically comes on when the mains lighting circuit fails. This means that when the building falls into darkness because of a power failure, the emergency lighting circuit takes over and a level of light is generated that is sufficient to enable people to leave in safety.

Also, emergency lighting must be provided if this will help shut down process machinery and so on prior to an evacuation.

Emergency lighting should be sited to cover specific areas

- intersections of corridors
- at each exit door
- near each staircase
- on each flight of stairs
- close to a change in floor level
- outside each final exit
- by exit and safety signs throughout the building
- within lift cars
- near firefighting equipment
- near each fire alarm call point.

Lighting units should be approximately 2 m off the floor – they need to be seen and could become obscured by smoke as smoke rises and accumulates below ceiling level.

Case study: Commercial buildings (2022)

In 2021, business safety officers from Devon and Somerset Fire and Rescue Service raised concerns with Mr *** for unsafe practices such as a lack of suitable means of escape in case of fire, structural failures and the absence of a fire risk assessment.

The officers took the decision that the premises were dangerous to the person who lived in the flat and, with immediate effect, they couldn't stay there anymore. This was because if a fire were to occur on the ground floor, the person in the flat above would not be able to make their escape safely. The person in the flat was also considered to be vulnerable as they were elderly with additional disabilities.

Following the investigation, the fire service made the decision to take Mr *** to court. The fire service determined that the conditions found at the premises in 2021 were so far below an acceptable standard that they had put people at risk of death or serious injury in the event of a fire occurring. The investigation found that Mr *** had known that the premises did not have adequate means of escape in case of fire back in 2009. He had never made an application for the works despite the need to do so. Despite this, he continued to take rent from a vulnerable person for several years despite knowing that the flat was unsafe to occupy.

After the prosecution, the Business Safety Legal Support Officer said

> Devon and Somerset Fire and Rescue Service take the safety of people who work, live and visit Devon and Somerset very seriously.
>
> Fire safety legislation is there for a reason, to ensure that people are safe in buildings should a fire occur and can safely evacuate should they need to do so.
>
> Where building owners provide accommodation, they must do so with all the required and appropriate fire safety measures in place.
>
> Where persons responsible for premises know that fire safety measures are poor and dangerous, or they knowingly allow poor fire safety standards, putting profits before safety, they will be held to account.

Mr *** pleaded guilty to all three charges from the RRO, including

- lack of compartmentation – fire separation failures between the businesses and the flat (Article 8).
- not having completed a fire risk assessment for the premises (Article 9)
- failing to provide suitable means of escape in case of fire from the flat (Article 14).

The fines awarded were the maximum allowed under the sentencing rules.

KEY RESOURCES

Regulatory Reform (Fire Safety) Order 2005

Building Regulations 2010 Fire Safety: Approved Document B (as amended June 2022)

BS 9999:2017 Fire safety in the design, management and use of buildings. Code of practice

BS 9991:2015 Fire safety in the design, management and use of residential buildings. Code of practice – under review 2023

Perry P
ISBN 978-0-7277-6647-2
https://doi.org/10.1680/fsbqa.66472.119

Chapter 8
Fire safety management in workplaces

Fire Risk Assessments have been carried out. What else must an employer do in respect of fire safety in the workplace?

Fire risk assessments are only one part of the fire safety responsibilities that employers have for the safety of their employees.

In addition, employers must

- provide and maintain such fire precautions as are necessary to safeguard those who use the workplace
- provide information, instruction and training to employees about the fire precautions
- nominate persons to undertake any special roles that are required under the emergency plan
- consult employees about the nomination of people to undertake special roles in respect of fire safety and about proposals for improving fire precautions
- inform other employers who share the workplace or have workplaces within the building about any of the significant fire risks identified that might affect the safety of their employees
- co-operate with other employers about any measures needed to reduce and control fire safety hazards and risks
- ensure that fire safety requirements are managed in any common parts and that the fire regulations are complied with. This may need to be undertaken by persons in control of the premises
- establish suitable means for contacting the emergency services and ensure that they can be called easily.

The Management of Health and Safety at Work Regulations 1999 requires employers to provide emergency plans that outline the procedures to be followed in emergencies. Fire would be one such emergency.

What should the emergency plan for fire safety cover?

The emergency plan must be in writing and readily accessible for those who might need it.

It should be site specific to the workplace and detail the pre-planned procedures in place for use in the event of a fire.

The emergency plan will usually include

- action on discovering a fire
- warning if there is a fire
- calling the fire brigade
- evacuation of the workplace
- procedures for those particularly at risk e.g. people with disabilities
- power or process isolation
- liaison with emergency services
- identification of key escape routes
- firefighting equipment provided
- specification of responsibilities in the event of a fire
- training required.

Emergency plans should be tested regularly.

If possible, involve the local fire authority.

What are some of the management controls that could be put in place for different aspects of fire safety?

The fire risk assessment will identify the control measures necessary to manage fire safety risks in the workplace.

Some of the main fire safety controls that an employer needs to manage are as follows

Emergency exit routes

Emergency exit routes are those that need to be clear and unobstructed to allow people to get out of the building quickly in an emergency.

They could be the same as those that people use to enter the building. But dependent on how big the building is and how far people are away from an exit, additional exits are often required.

Escape routes must lead to a place of safety.

Check that fire escape routes are clear and unobstructed at all times. Make sure that they are easily accessible. Check that the area into which the escape route leads is clear and not obstructed. Can doors open outwards or are they blocked by cars, bins, barrels and so on?

Check that any emergency lighting works.

The following items should not be stored on protected fire escape routes or on any route that gives emergency exit from a building

- portable heaters of any type
- heaters that have unprotected naked flames or radiant bars
- fixed heaters using a gas supply cylinder, where the cylinder is in the escape route
- oil-fuelled heaters or boilers
- cooking appliances
- upholstered furniture
- coat racks
- temporarily stored items – for example, furniture, packaging and deliveries
- lighting using naked flames
- gas boilers, pipes and meters (except those meeting building regulation standards)
- gaming or vending machines
- electrical equipment other than normal lighting and emergency lighting.

Fire doors

Fire doors hold back smoke and flames. They should always be kept closed. Many fire doors are self-closing – do *not* prop them open.

Fire doors usually have 30 min fire protection – they will hold back the flames and heat for 30 min before beginning to burn so they keep fire escape routes safe for a short period of time.

Fire doors also hold back smoke if they are close-fitting enough.

Fire doors should open in the direction of travel and be easy to use and either have a push bar or push pad for easy opening.

Fire doors must *never* be locked. If you need to keep a door locked for security reasons, it must have a break glass bolt or lock so that it can be opened in an emergency.

Fire safety signage

Fire safety signage should comply with the RRO and the Health and Safety (Safety Signs and Signals) Regulations 1996. Where possible, fire signage should comply with the relevant British Standards or International Codes, notably BS EN ISO 7010:2020 as the current standard and BS 5499 as an older standard.

While there is no requirement to change old BS 5499 signs to the new BS EN ISO 7010 signs, you should not mix the two on any site. If you do need to install new signs, it is necessary to ensure that the existing signs are the same type.

Whatever format of signs are used in a premises, the format must be consistent throughout the building to avoid confusion for employees and visitors.

Signs help people know what to do and where to go when they are unfamiliar with premises and locations.

All premises *must* have clearly visible fire safety signs that

- indicate fire exit routes
- identify fire doors
- identify fire alarm call points
- identify firefighting equipment.

Fire signs must now include a pictogram and a directional arrow. Words or text signs are no longer legal. Pictograms are symbols that should be understood by everyone, no matter what language they speak.

Signs must be clearly visible so need to be of the correct size.

Some fire exit signs must be illuminated – usually those on the final exit doors.

Even though you do not need to install illuminated signs, it is a good idea to do so as it aids emergency evacuation in poor light or darkness.

An alternative provision is to display photoluminescent signs. These glow in the dark and are much more clearly seen.

Fire signs must be of the right colour as follows.

Fire exit signs	Green background White writing
Assembly point	Green background White writing
Fire doors – keep shut or locked	Blue background White writing
Fire call points Fire extinguishers }	Red background White writing

What is 'wayfinding' signage?

The Fire Safety (England) Regulations 2022 made it a legal requirement from 23 January 2023 for all high-rise residential buildings – that is, those that are at least 18m in height or more than seven storeys – in England to install wayfinding signage in their buildings. This includes clear markings identifying floor and individual flat numbers.

The Grenfell Tower Inquiry recommended that wayfinding signage be improved in existing high-rise buildings. The government decided to introduce a legal requirement for wayfinding signage and included a clause in the new Fire Safety (England) Regulations 2022. They considered that in high-rise buildings the risk of firefighters becoming disorientated in smoky conditions was greater than in smaller buildings and did not consider requiring the signage to be retrofitted in all existing multi-occupied residential buildings.

It is a legal requirement for all new multi-occupied residential buildings above 11 m (including those existing buildings undergoing relevant material alterations) to have signage installed as a requirement of the Building Regulations.

Signage must indicate which floor the individual is on, and all flats and apartments must have clear numbering visible to those accessing corridors, common areas, vestibules and so on.

Signs must be visible in dark and smoky conditions and meet the following standards.

- Signs should be located on every landing of a protected stairway and every protected corridor/lobby (or open access balcony) into which a firefighting lift opens.
- Text should be in sans serif typeface with a letter height of at least 50 mm. The height of the numeral that designates the floor number should be at least 75 mm.
- Signs should be visible from the top step of a firefighting stair and, where possible, from inside a firefighting lift when the lift car doors open.
- Signs should be mounted between 1.7 m and 2 m above floor level and, as far as practicable, all the signs should be mounted at the same height.
- Text should be on a contrasting background, easily legible and readable in low level lighting conditions or when illuminated with a torch.

What should I put where?

	A general sign to indicate a fire exit.
	A fire exit that leads downwards or straight on.
	The fire exit is to the left of the sign.
	The fire exit is to the right of the sign – suitable for all areas.
	General fire exit sign – can be displayed above main entrance doors.
	General fire exit that is to the left of the sign – suitable for all areas.
	General fire exit that is to the right of the sign – suitable for all areas.
	The fire exit is to the left of the sign – suitable for all areas.
	Indicates that the fire door has a push bar to open and that it is a safe area to be using.

	The fire exit is to the right of the sign – suitable for all areas.
	Add this sign if the fire exit route is at an angle and leads upwards – for example, upstairs from a basement
	Add this sign to a 'word' sign to make the word sign legal. Fire exit is to the left.
	Exit towards the left for the nearest fire exit door.
	Go downstairs to the nearest fire exit route.
	Exit towards the right for the nearest fire exit door.
 	The 'running man' symbols can be displayed next to text or word signs to make them legal.
 	Text and pictogram general fire exit signs, indicating whether to go to the right or left.

	A sign which usually is displayed next to the fire extinguishers or where the fire alarm is.
	A sign that indicates the fire alarm call point is nearby.
	A fire alarm call point sign.
	Usually displayed adjacent to water or some foam fire extinguishers.
	A general sign displayed next to fire extinguishers. *Note that this sign should be used for all those indicating fire extinguishers.*
	Displayed next to carbon dioxide fire extinguishers.
	A fire hose reel sign, usually displayed next to the fire hose.
	A telephone kept especially for raising the fire alarm.

Any of these signs must be on doors which must be kept shut or locked for fire safety reasons – for example

- electrical cupboards
- lift motor rooms
- chemical storage cupboards
- all fire exit doors where they have self-closing devices.

Note that fire exit route doors must never be kept locked shut.

Fire action signs

Signs displaying the actions you want people to take in case of a fire need to be displayed in suitable locations around the premises. Usually, they are displayed near fire call points and adjacent to emergency exits and/or fire extinguishers.

Fire action signs are simple and need important pieces of information added.

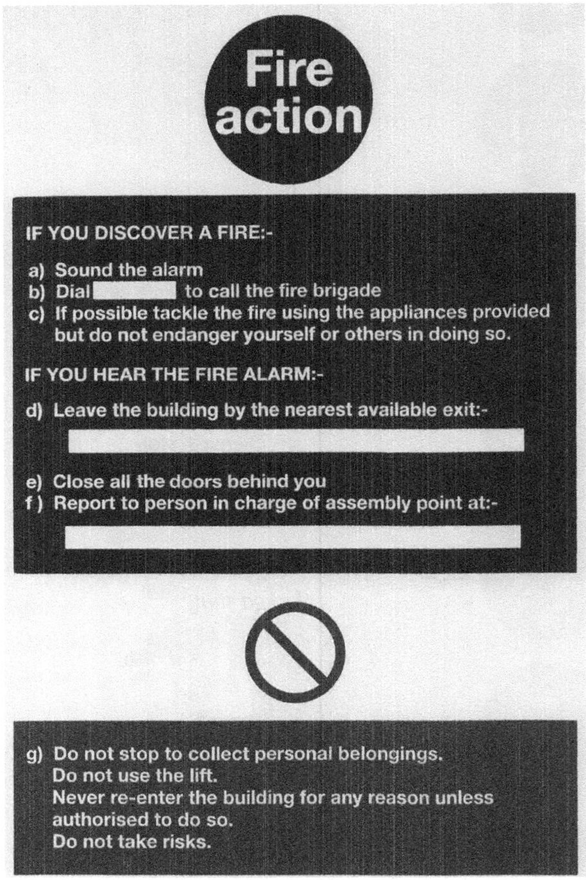

Assembly point

You need to designate an area where staff and others can report that they are safe. This could be the car park if you have one, or a building across the road. Someone needs to be 'in charge' so that when the fire brigade arrives, they can tell them whether anyone is left inside the building or missing.

Summary

- Fire action notices are mandatory for all premises.
- Fire exit signs are needed for all but the smallest and simplest of properties.
- Fire extinguisher identification signs and extinguishers are mandatory; all premises must have at least one extinguisher identification sign.
- Fire alarm call point signs are mandatory if the building has a fire alarm.
- Other fire equipment signs are used to signpost other information – for example, hose reels or dry risers warning and prohibition signs will be required where extra risk of fire exists on a premises.

Evacuation procedures

If the fire alarm is raised or there is a fire in the building, you *must* evacuate people quickly and safely.

All staff must be trained in what to do in an emergency – that is, who goes to check the toilets, work areas, store rooms, customer areas and so on. Choose someone to be responsible and to take charge of an incident. This person is often called the fire warden.

People must be asked to leave calmly and without panic.

Do not wait until you have checked for a fire or seen how 'real' the emergency is – leave at once.

Guide people to the emergency exit routes. Make sure all areas of the building are checked and do not forget plant rooms.

All staff should be accounted for at the assembly point.

Do not use any lifts.

Make sure someone has called the fire brigade.

If possible, close all doors and windows, but do not take any risks to anyone's safety.

Remember that disabled customers may have difficulty getting out quickly, so pay particular attention to their needs.

Fire drills are a good thing to do for training purposes, although difficult to do with customers. Try to run through the fire drill and evacuation procedure regularly with staff, and there is no reason not to have a fire drill before opening times. Keep a record of when and who took part.

Top tips for emergency exit routes

- keep fire exit routes clear of obstructions at all times
- check them regularly throughout the day
- keep fire doors shut
- check that door seals are intact
- smoke usually kills people, not flames, so make sure that smoke cannot travel around the building
- *never* lock fire doors – use a break glass bolt if you need security
- obstructed fire exit routes can cause you to be closed down
- you do not get a second chance to put it right – get it right first time.

Top tips for fire safety signage

- designate an assembly point
- nominate someone to be in charge of an incident
- check that your fire signs can be clearly seen
- buy photoluminescent signs as these show up in the dark
- make sure that the details on the fire action notice (see above) are completed and that these are displayed at exit routes and at fire call points.

Top tips for evacuation procedures

- have a practice fire drill
- make sure staff know what to do to evacuate the premises
- nominate someone to check all areas of the premises – for example, plant rooms
- keep emergency routes clear of obstructions and well visible
- choose an assembly point and check staff off against the register
- do not forget other contractors and workmen who may be working in the premises
- appoint a 'fire warden', someone to take extra responsibility for making sure that you and your staff are 'fire aware' and who takes charge of any incident.

How often should I carry out a fire safety check in my premises?

Regular fire safety checks reduce the risks from fire, prevent complacency and identify shortcomings in any of the control procedures in place.

The Management of Health and Safety at Work Regulations 1999 require employers to control, monitor and review any arrangements or preventative measures implemented in the workplace in order to control hazards and risks.

Fire safety controls can easily be abused – for example, fire exit routes become blocked, fire extinguishers may be empty or the fire alarm may not work.

A systematic and regular procedure for fire safety checks will help prevent deterioration in control measures.

The fire risk assessment will identify the control measures necessary to reduce or eliminate the hazards from fire and during the risk assessment process, a consideration of how frequently those controls need to be checked should be made.

Fire safety checks should become second nature and a quick visual check of the workplace should be carried out daily, before work starts.

If employees have had training in fire safety, they will be familiar with recognising hazards and should be encouraged to play an active part in managing fire safety in the workplace.

A visual check of fire exit routes every day will soon identify whether they are obstructed. If they are, steps can be taken then and there to remedy the defect.

More detailed fire safety checks could be undertaken monthly by using a fire safety checklist and recording the findings.

As with any control system for health and safety, the employer will need to show that they were aware of the hazards and risks and had an effective control procedure in place. An effective control procedure can only be demonstrated if it is regularly reviewed and shown to be working.

What subjects and areas should be covered in a fire safety check?

A fire safety check should be an objective review of the fire safety hazards, risks and precautions associated with the workplace.

The fire safety check should be relevant to the hazards and risks associated with the workplace and should not be a long list of every eventuality in respect of fire risks.

It is probably best practice to develop individual company fire safety checklists. There are examples available in various publications, but it shouldn't take too long to devise a bespoke form.

A fire safety checklist should include

- good housekeeping
    - areas free of rubbish
    - rubbish removed frequently
    - no excess and combustible materials
    - personal workplaces tidy

- storage areas
  - what is stored where
  - removal of refuse
  - safe storage of items
  - room to move around piles of stock
  - no stock obstructing vents and so on
  - no stock obstructing sprinkler heads or fire alarm systems
  - storage areas kept separate from work areas, doors kept shut and so on
- smoking
  - is a no-smoking policy in place
  - are smokers materials dealt with safely – for example, ashtrays emptied into metal containers
  - are smoking rooms checked frequently
  - are fire detection devices working?
- flammable substances
  - any stored in the work area
  - are they needed
  - kept away from ignition sources
  - used by trained people
  - consideration given to fumes and vapours
  - ventilation grilles unobstructed
  - metal, non-combustible storage boxes
  - used with 'permit' procedure
  - lids on containers and so on
- equipment maintenance
  - routine maintenance
  - no 'friction' burns
  - machinery clean
  - no build-up of grease
  - no oily rags kept near equipment
  - no combustible materials nearby
- flammable gases
  - any use of LPG
  - stored externally

    - separated area
    - used with 'permit' procedure
    - empty cylinders stored safely
    - regular checks

- heating and lighting

    - heating appliances correctly used and positioned
    - no use of employees' heaters
    - regular maintenance
    - no combustible materials in the vicinity
    - correct light bulb wattages
    - no combustible lamp shades
    - nothing stored under lights or other radiant heat

- electrical appliances

    - used correctly
    - tested regularly
    - maintained properly
    - visual checks daily
    - correct fuses, plugs and so on
    - safety shut off switches accessible
    - safety devices working

- arson

    - good security systems
    - regular parameter checks
    - combustible materials locked away
    - no available ignition sources

- fire protection systems

    - fire alarms working
    - firefighting equipment available
    - all equipment regularly checked
    - sprinklers checked
    - smoke and heat detectors in place
    - no obstructions

- means of escape

    - clear and accessible from all areas
    - unobstructed routes

  - doors kept shut
  - panic bolts working
  - no locked fire doors
  - suitable lighting
  - emergency lighting working
  - signage displayed clearly
  - unambiguous instructions
  - identification of assembly point
  - fire action notices displayed

- staff training

  - has everyone had fire safety induction training?
  - fire wardens appointed and trained
  - contractors given appropriate information of fire procedures
  - visitors given information

- emergency procedures

  - who does what and when
  - raising the alarm
  - calling the fire brigade
  - meeting the fire brigade
  - searching the building
  - fighting the fire

- management procedures

  - completion of fire risk assessments
  - review of risk assessments
  - availability of emergency plan
  - competent persons
  - monitoring and review processes
  - coordination of departments regarding fire hazards
  - use of permit to work and hot works permits
  - control of contractors
  - staff training programme.

The above are some of the key subjects that can form the basis of a fire safety checklist. The format can be questions and yes/no answers, or statements and satisfactory/unsatisfactory answers. Any fire safety checklist, however, should contain a section that identifies any shortcomings in the control measures needed, with remedial action to be taken, by whom and by when. The findings of the fire safety check should be reviewed to ensure that remedial works have been completed.

FIRE SAFETY CHECKLIST

A regular fire safety check of the premises should be carried out. Complete this checklist monthly and keep a record. You should also make a daily pre-opening check that fire safety measures are satisfactory.

Fire safety management	**Yes**	**No**	**Not applicable**
1. Do you carry out daily checks on all fire safety measures?			
2. Are staff trained in emergency procedures?			
3. Have you got an emergency plan?			
4. Have you completed fire risk assessments and/or updated earlier versions?			
5. Is the fire certificate available at all times?			
6. Do you comply with conditions of any special licences?			
Means of escape			
7. Are all escape routes clear of obstructions?			
8. Are doors kept shut on escape routes?			
9. Are all internal fire doors clearly labelled?			
10. Can all fire safety signs and call point signs be seen clearly?			
11. Are any fire doors locked or not able to be opened?			
12. Are corridors used for any storage of combustible materials etc?			
Fire alarm			
13. Is the fire alarm tested weekly and records kept?			
14. If a manual alarm is used, is it tested and easily available?			
Emergency lighting			
15. Is it tested regularly?			
16. Are records kept of tests?			
Other detection equipment			
17. Are regular tests carried out and records kept?			

Fire instructions			
18. Do all staff know what to do in the event of a fire?			
19. Are fire action notices displayed?			
20. Is an assembly point designated?			
Housekeeping			
21. Is refuse removed regularly?			
22. Is combustible material kept away from heat sources?			
23. Are aerosol cans stored safely?			
24. Are boiler rooms and electrical cupboards kept free from combustible materials?			
Electrical			
25. Are sockets overloaded?			
26. Is 'one plug one socket' rule followed?			
27. Have checks been carried out on electrical plugs, leads and appliances?			
28. Is equipment earthed?			
Smoking			
29. Are cigarette ends properly discarded into metal receptacles?			
30. Is furnishing non-combustible?			
31. Are staff trained on what to look out for regarding smoking material and smouldering fires etc?			
32. Do staff follow 'no smoking' rules?			
Heating			
33. If LPG heaters are used are they serviced regularly?			
34. Are LPG heaters switched on by trained people?			
35. Are LPG cylinders stored outside the premises?			
36. Is combustible material kept away from heaters?			
37. If open fires are used are they monitored, with fire guards?			
38. Are heaters securely fixed?			

Lighting			
39. Are equipment, combustible materials etc. kept away from light bulbs?			
40. Is the correct light bulb wattage used for the lampshade?			
Flammable substances			
41. Are substances kept away from heat sources and in metal containers?			
42. Are cleaning materials correctly stored away from heat?			
43. Are chemicals used safely and not near naked flames?			
Storage			
44. Are storage areas easy to get to, are electrical sockets visible and is stock away from any heat source?			
Kitchens			
45. Are canopies cleaned regularly?			
46. Are deep fat fryers thermostatically controlled?			
47. Are grease deposits cleaned regularly?			
48. If heat lamps/blow lamps are used for flambé dishes, is there a risk assessment?			
49. Are thermostats working and maintained?			
50. Are electrical and gas equipment services regularly?			
Night checks			
51. Is there a proper procedure for checking the premises at night?			
Other			
52. Are contractors' hot works controlled?			
53. Are contractors' works checked?			
54. Are all employees fully trained in fire safety?			
55. Is there any activity in the vicinity that could cause an increased fire risk to the premises?			

Are there any guidelines that indicate how often different types of fire safety equipment need to be checked?

Guidelines on the frequency of fire safety checks for equipment are contained in the publication 'Fire safety in the workplace' issued by the government.

In addition, the Loss Prevention Council and Fire Protection Association issue guidance on the testing and maintenance of fire safety equipment.

Manufacturers' instructions on maintenance must also always be followed.

The following provides a summary of weekly checks.

Fire detection – for example, automatic detectors, smoke alarms and heat detectors	Weekly test to ensure that they are working. Weekly visual inspection.
Fire alarm systems	Weekly test to ensure that they are working. Weekly check of call points.
Emergency lighting	Weekly check of all manual systems – for example, torches. Visual check of all light fittings.
Fire extinguishers	Weekly check to ensure all extinguishers are in the correct location. Weekly check to ensure that all are full.

Monthly checks of all fire safety equipment should test that

- all equipment is working
- emergency call points operate
- all emergency lighting works
- smoke and heat alarms work
- firefighting equipment is in place.

Annual checks should be carried out by a competent person and in accordance with manufacturers' instructions. A full check and test of the entire system should be carried out by service engineers.

All checks, whether weekly, monthly or annual, must be recorded.

The following record sheets are examples of the types of forms that can be used.

FIRE EXTINGUISHER AND HOSE REELS – RECORD OF TESTS AND INSPECTION

- Check all equipment for correct installation and function on a weekly basis.
- Arrange for annual servicing and testing by a competent engineer.

Date	Type	Location	Tests or inspection	Satisfactory Yes/No	Remedial action taken and signature

FIRE ALARM SYSTEM – RECORD OF TESTS

- Check for state of repair and operation on a weekly basis.
- Arrange for annual servicing and testing by a competent engineer.

Date	Call point number	Pass/fail	Action required	Signed

EMERGENCY LIGHTING SYSTEM – RECORD OF MAINTENANCE AND TESTS

- Check all lights for state of repair and correct functioning on a weekly basis.
- Arrange for annual servicing and testing of the system by a competent engineer.

Date	Satisfactory Yes/No	Type of defect and location	Remedial action taken and signature

FIRE DETECTION SYSTEM – RECORD OF TESTS

- Arrange for annual servicing and testing by a competent engineer.

Date	Type of detectors: Smoke/heat	Location	Detector activated	Satisfactory Yes/No	Action required	Signed

SUPPRESSION SYSTEMS – RECORD OF TESTS

- Arrange for annual servicing and testing by a competent engineer.
- Use this form to record tests on any sprinkler system or Ansul system, or any other fire suppression system installed (if any).

Date	Type of system	Location	Detector activated	Satisfactory Yes/No	Action required	Signed

Are there any structural features of a building that can be checked to ensure fire safety?

The most serious feature that increases the risk of fire spreading throughout a building is unprotected openings in compartment walls, floor and ceilings.

Fire safety design works on the principle of compartmentation, which ensures that if a fire starts in an area it cannot spread beyond that area.

Any compromise to compartmentation will allow heat, fumes, smoke and toxic fumes to spread rapidly around a building.

Structural features to check are

- ducts without dampers
- flues
- redundant chimneys
- voids behind panelling, floors, ceilings
- holes around service pipes
- uncompartmentalised roof voids
- unprotected staircases
- warped and ill-fitting doors
- missing or removed doors
- removed partition walls.

It may be necessary to consult architects' drawings, design specifications or the mechanical and electrical manuals and drawings.

Repairs and refurbishment works often create fire safety hazards by breaching compartmentation, and the contractors often fail to reinstate the integrity of the wall, ceiling and so on.

When refurbishment works, new air conditioning and so on are to be undertaken, it is imperative to agree remedial actions. Architectural advice should be sought, but for those projects undertaken in house, the project manager should have regard to the likely breach of structural fire safety measures.

External cladding must now be checked on a regular basis.

What is a 'hot works permit'?

A hot works permit is a safety management procedure for controlling the use of equipment and processes that use heat generating equipment and which, therefore, pose a risk of fire. It is a formal written system.

A hot works permit is issued for every job that uses hot work, heat generating equipment or heat generating substances.

A hot works permit records the safety management procedures in respect of certain job activities.

Any activity involving

- blow torches
- hot bitumen boilers
- welding
- use of LPG
- use of flammable substances

should be accompanied by a hot works permit.

What information should a hot works permit contain?

The permit to work form must help communication between everyone involved in carrying out the task, or who will be affected by the task.

The main elements of a permit system are

- permit title
- job location
- identification of plant to be worked on
- description of work to be done
- equipment to be used
- hazard identification
- control measures necessary
- protective equipment/clothing to be used
- training and competency of individuals
- identification of who will be carrying out the tasks
- authorisation by competent person
- handback procedure
- acceptance of safety procedures
- final checks on handbook
- signatures
- site safety rules.

Each permit should be designed for the specific job and should address the hazards and risks and outline the controls to be taken to eliminate/reduce risks.

Permits should be kept for a reasonable period, and it may be necessary to review the safety precautions taken on a specific job.

Key topics summary

- all fire safety signs should clearly and effectively mark escape routes and exits
- fire doors and escape routes must be clearly marked
- signage must be displayed along an exit route
- all fire safety signs should be illuminated
- escape route signs should feature direction arrows
- emergency exit signs above fire doors should not have arrows
- all fire escape route signs should be in picture form
- braille and tactile fire safety signage should also be displayed
- signage should be visible from a distance and at an appropriate height
- employees should be trained in the use of fire extinguishers
- employees should be aware of the location of fire alarms
- employees should be trained in what to do in the event of fire
- all firefighting equipment must be clearly identified.

KEY RESOURCES

Regulatory Reform (Fire Safety) Order 2005
Fire Safety Act 2021
Fire Safety (England) Regulations 2022
Management of Health and Safety at Work Regulations 1999
Health and Safety at Work etc. Act 1974
Health and Safety (Safety Signs and Signals) Regulations 1996
Safety signs and signals. The Health and Safety Regulations 1996. Guidance on Regulations L64: 2015 (3rd edition)
Building Regulations 2010 Fire Safety: Approved Document B (as amended 2020)
BS EN ISO 7010:2020 Graphical symbols. Safety colours and safety signs

Perry P
ISBN 978-0-7277-6647-2
https://doi.org/10.1680/fsbqa.66472.147

Chapter 9
Hazardous and flammable substances

What are flammable liquids?

Flammable liquids give off large volumes of flammable vapours at room temperature.

These vapours when mixed with air, can ignite, often violently.

Spilled flammable liquids can flow considerable distances to an ignition source and the source of ignition could be in a quite different area from the spilled liquid.

Spills on clothing of flammable liquids create considerable fire risks.

Flammable liquids are therefore considerable fire safety hazards and need to be closely managed and controlled to eliminate or reduce their contribution to a fire.

A flammable liquid may be reasonably safe in a stored *lidded* container but if the lid is poorly fitting, fumes can escape and be potential for ignition.

What controls need to be implemented for flammable liquids?

The following guidelines should be followed when storing *any* flammable liquids

- avoid storing flammable liquids if possible
- substitute the product for something less flammable
- store flammable liquids in a separate storage area or in a purpose-made, fireproof bin or cupboard
- dispense and use them in a safe place, where there is good ventilation and no heat or ignition sources
- keep containers closed when not in use and check lids regularly
- use safety containers with self-closing lids if possible
- dispense liquids over a tray or shallow container to contain any spillage to a localised area
- keep non-flammable, absorbent material close to hand to deal with spillages – for example, sand
- clear up spillages immediately
- dispose of contaminated rags and materials safely and remember that these have become flammable substances. Dispose of these separately and in flameproof containers. Do not store quantities of contaminated materials together as they may 'spontaneously' combust due to heat generation and ignite the vapours given off of the materials.

What are flammable dusts?

Fine particles of dust in certain circumstances can accumulate and explode, causing both physical damage and potential fire risks.

Flammable dusts may arise from solid materials that are flammable or that may contain flammable substances or components and any accumulation that has the potential to ignite.

The following controls in relation to dusts are required

- keep dust quantities to a minimum and prevent escape to the atmosphere
- use local exhaust ventilation to remove dusts at source
- regularly empty dust bags so as to reduce the volumes collected
- keep machinery clean
- clean the workplace by regular vacuuming and so on
- clear up spillages as they occur
- follow any inherent dust control procedures and safety procedures associated with the plant.

What are flammable solids?

Some types of 'solid' material will ignite easily and burst into flames with a fierce and hot burning intensity.

Examples of flammable materials

Flammable liquids

- petroleum products
- petrol, diesel
- cooking oils
- motor oils
- hydraulic fluids
- lubricants
- solvents
- paints
- thinners
- degreasing agents.

Flammable dusts

- flour
- confectionary dusts
- swarf
- wood dusts.

Flammable substances/solids

- foam
- polystyrene
- textiles
- polyester wadding
- building products – for example, insulation materials.

Safety procedures to follow would be

- do not store flammable materials close to any heat source – for example, heaters, fans, cookers and electrical equipment
- illuminate all potential sources of ignition
- store small quantities of materials
- do not use highly flammable materials in high-risk fire areas
- keep all passageways, work areas and so on clear so that if a fire were to start it cannot 'leap' to adjoining fuel sources
- maintain good general fire precautions.

What are flammable gases?

Flammable gases are often stored in cylinders under pressure and any uncontrolled release of the contents can cause explosion and fire. They gases will be those of the liquefied petroleum gas family – for example, methane, acetylene, butane and propane.

Safety procedures are

- store cylinders of gas safely and securely
- store cylinders in an open-air secure cage, compound
- secure all valves in the 'shut' position
- protect cylinders from impact damage
- use and fit the correct equipment – for example, pressure valves, hoses and connectors
- check hoses, valves and so on regularly
- do not use faulty equipment
- do not tamper with valves unless trained and competent to do so.

Why is oxygen a fire hazard?

In certain circumstances, known as an oxygen-enriched atmosphere, oxygen will fuel a fire rapidly and will contribute to the risk of explosion. Materials that will ordinarily burn slowly may burn very vigorously in an oxygen-enriched atmosphere.

Some substances may burst into flames in an oxygen-enriched atmosphere – for example, greases and oils.

Safety procedures to follow include

- never use oxygen in place of compressed air
- never use oxygen to 'sweeten' the air in a confined space or working area – for example, tanks and vats
- never use grease or oil on equipment containing oxygen.

What is an oxidising substance?

An oxidising substance is one that can cause a flammable material to ignite because it releases oxygen and increases the potential for combustion.

Oxidising substances must carry a hazard warning sign and must not be stored adjacent to flammable materials.

As an employer, I need to store flammable substances in my workplace. What safety measures do I need to take?

The measures stated in the previous questions must be followed depending on what type of flammable substance you have.

In addition, as an employer you must

- make sure that workers know about the hazards and risks of these substances
- ensure that employees are aware of the risk assessments
- ensure that employees have had training on how to use flammable substances, how to store them safely and so on
- ensure that employees are aware of the dangers of mixing chemicals, liquids and substances
- introduce permit to work and hot works permits
- introduce and monitor 'safe systems of work'
- train all employees in what to do in an emergency
- increase any firefighting provisions to address the increased risk
- apply for a special fire certificate if you store above certain limits of flammable substances
- manage and maintain procedures for dealing with spillages
- consider the need for any special first aid equipment and specialist advice for treating injuries.

What are the Dangerous Substances and Explosive Atmospheres Regulations 2002 about?

The Dangerous Substances and Explosive Atmospheres Regulations 2002 (DSEAR) (as amended) is a set of regulations concerned with the protection against risks from fire, explosion and similar events arising from dangerous substances used or present within the workplace.

The regulations apply to employers and the self-employed.

A dangerous substance is one which, because of its properties or how it is used, could cause harm to people from fires and explosions.

Dangerous substances include

- petrol
- liquified petroleum gas
- paints
- varnishes
- solvents
- dusts which when mixed with air could cause an explosive atmosphere.

Dangerous substances are found in some quantity in most workplaces – for example, paints and thinners, even if stored in a maintenance shed or an office.

An explosive atmosphere is an accumulation of gas, mist, dust or vapour which, when mixed with air, has the potential to catch fire or explode.

While the explosive atmosphere may not always explode, if a fire were to start it would travel rapidly through the atmosphere, increasing in intensity – for example, in a confined space. The rapid spread of flames or rise in pressure could cause an explosion.

What does an employer need to do under the DSEAR?

As an employer, you must

- carry out a risk assessment of any work activity involving dangerous substances
- provide measures to eliminate or reduce risks as far as is reasonably practicable
- provide equipment and procedures to deal with accidents and emergencies
- classify places where explosive atmospheres may occur in 'zones' and mark the zones as necessary
- provide information and training to employees.

The important part of the regulations is to carry out a risk assessment of the use and storage of dangerous substances.

Any fire risk assessment should consider the need to manage the risks from substances likely to contribute to the risks of fire.

Emergency evacuation procedures may need to be reviewed because the spread of fire may be rapid, and it will be vital to ensure that people can evacuate the building quickly.

How will I know if a substance is a dangerous substance under the DSEAR?

The first step is to check whether the substances have been classified under the Chemicals (Hazard Information and Packaging for Supply) Regulations 2002 (CHIP) as

- explosive
- oxidising
- extremely flammable
- highly flammable
- flammable.

The substance must carry one of the hazard warning symbols – an orange square with a black symbol.

Substances must be provided with safety data sheets. These contain all the chemical, physical and hazard safety information regarding the product and must be considered when determining whether substances fall within the remit of the regulations.

Some substances may not carry a hazard warning sign but the way in which they are used in the work environment may cause them to be dangerous or highly explosive.

It is a combination of the substance and the circumstances in which it is used which is important.

Diesel oils, for example, are not classified as flammable under CHIP regulations yet, if they are heated to a sufficiently high temperature, they can create a fire risk.

KEY RESOURCES

Control of Substances Hazardous to Health Regulations 2002
Dangerous Substances and Explosive Atmospheres Regulations 2002
Regulatory Reform (Fire Safety) Order 2005
Dangerous Substances and Explosive Atmospheres Regulations 2002. Approved Code of Practice and guidance (second edition 2013)
Chemicals (Hazard Information and Packaging for Supply) Regulations 2002 (CHIP)

Perry P
ISBN 978-0-7277-6647-2
https://doi.org/10.1680/fsbqa.66472.153

Chapter 10
Fire safety in high-rise residential buildings

What caused the Grenfell Tower fire in 2017 and why did it have such tragic consequences?

The public inquiry into the Grenfell Fire concluded

> The fire was started by an electrical fault in a large fridge-freezer in the kitchen of Flat 16.
>
> The fire took hold in the external cladding of the tower block. It is most likely to have entered the cladding as a result of hot smoke impinging on the uPVC window jamb, causing it to deform and collapse and thereby provide an opening into the cavity between the insulation and the aluminium composite material (ACM) cladding panels through which flames and hot gases could pass.
>
> Once the fire had escaped from Flat 16, it spread rapidly up the east face of the tower. It then spread around the top of the building in both directions and down the sides until the advancing flame fronts converged on the west face near the south-west corner, enveloping the entire building in under three hours.
>
> The principal reason why the flames spread so rapidly up, down and around the building was the presence of the ACM rainscreen panels with polyethylene cores, which acted as a source of fuel.
>
> The principal mechanism for the spread of the fire horizontally and downwards was the melting and dripping of burning polyethylene from the crown and from the spandrel and column panels, which ignited fires lower down the building. Those fires then travelled back up the building, thereby allowing the flame front to progress diagonally across each face of the tower.
>
> The presence of polyisocyanurate and phenolic foam insulation boards behind the ACM panels, and perhaps components of the window surrounds, contributed to the rate and extent of vertical flame spread.

The rapid spread of the fire was caused by combustible materials and the subsequent failure of compartmentation within the building.

Residents of the tower block were quickly affected by smoke, toxic fumes and heat. Many fire prevention measures were missing or failed within the building and the means of escape procedures were not readily known by residents. Star lobbies quickly become full of smoke and escape was almost impossible.

The common fire escape plan in residential tower blocks was a 'stay put' policy whereby residents were to stay in their flats until advised by the fire brigade that it was safe to leave. Many residents died because they followed this policy, including when given advice by the emergency services when they dialled 999.

What are some of the lessons relating to high-rise buildings that have been learned so far from the Grenfell fire?

The Phase 1 report of the public inquiry, led by Sir Martin Moore-Bick, contained a number of recommendations that the government have committed to adopting in full.

In summary, the key recommendations and learnings are

- the owner and manager of every high-rise residential building be required by law to provide their local fire and rescue service with information about the design of its external walls together with details of the materials of which they are constructed and to inform the fire and rescue service of any material changes made to them
- all fire and rescue services ensure that their personnel at all levels understand the risk of fire taking hold in the external walls of high-rise buildings and know how to recognise it when it occurs
- the owner and manager of every high-rise residential building be required by law to provide their local fire and rescue services with up-to-date plans in both paper and electronic form of every floor of the building, identifying the location of key fire safety systems
- the building contains a premises information box, the contents of which must include a copy of the up-to-date floor plans and information about the nature of any lift intended for use by the fire and rescue services
- all fire and rescue services be equipped to receive and store electronic plans and to make them available to incident commanders and control room managers
- the owner and manager of every high-rise residential building be required by law to carry out regular inspections of any lifts that are designed to be used by firefighters in an emergency and to report the results of such inspections to their local fire and rescue service at monthly intervals
- the owner and manager of every high-rise residential building be required by law to carry out regular tests of the mechanism that allows firefighters to take control of the lifts and to inform their local fire and rescue service at monthly intervals that they have done so
- the owner and manager of every high-rise residential building be required by law to draw up and keep under regular review evacuation plans, copies of which are to be provided in electronic and paper form to their local fire and rescue service and placed in an information box on the premises

- all high-rise residential buildings (both those already in existence and those built in the future) be equipped with facilities for use by the fire and rescue services enabling them to send an evacuation signal to the whole or a selected part of the building by means of sounders or similar devices
- the owner and manager of every high-rise residential building be required by law to prepare personal emergency evacuation plans (PEEPs) for all residents whose ability to self-evacuate may be compromised (such as persons with reduced mobility or cognition)
- the owner and manager of every high-rise residential building be required by law to include up-to-date information about persons with reduced mobility and their associated PEEPs in the premises' information box
- in all high-rise buildings floor numbers be clearly marked on each landing within the stairways and in a prominent place in all lobbies in such a way as to be visible both in normal conditions and in low lighting or smoky conditions
- the owner and manager of every residential building containing separate dwellings (whether or not it is a high-rise building) be required by law to provide fire safety instructions (including instructions for evacuation) in a form that the occupants of the building can reasonably be expected to understand, taking into account the nature of the building and their knowledge of the occupants
- the owner and manager of every residential building containing separate dwellings (whether or not they are high-rise buildings) be required by law to carry out an urgent inspection of all fire doors to ensure that they comply with applicable legislative standards
- the owner and manager of every residential building containing separate dwellings (whether or not they are high-rise buildings) be required by law to carry out checks at not less than three-monthly intervals to ensure that all fire doors are fitted with effective self-closing devices in working order
- all those who have responsibility in whatever capacity for the condition of the entrance doors to individual flats in high-rise residential buildings, whose external walls incorporate unsafe cladding, be required by law to ensure that such doors comply with current standards.

The Phase 1 report contained several additional recommendations addressing fire and rescue procedures, command control systems and communication, firefighter training and other matters.

How does the Grenfell fire affect fire risk assessments for residential buildings, especially those that are high-rise?

The Regulatory Reform (Fire Safety) Order 2005 (RRO) does not apply to domestic premises unless part of the premises is used as a workplace.

The RRO does, however, apply to the common parts of any residential building – that is, staircases, lobbies, corridors, plant rooms, roof terraces and external areas.

The responsible person – that is, the owner, landlord or managing agent for the residential building – must carry out a fire risk assessment of the common parts.

The Grenfell fire has strengthened the need for fire risk assessments of the common parts and has identified the need to assess external cladding and other building materials as part of the fire risk assessment.

More emphasis needs to be put on the integrity of the structural elements of a building and these should be included in the fire risk assessment.

Fire stopping measures need to be checked to ensure that they have not been breached, especially where refurbishment works have been carried out.

The Grenfell Tower Inquiry has identified that the door to individual accommodation must also be regularly assessed to ensure that it has the correct standard of fire protection – that is, they are effective fire doors. Many residents, whether owners, leaseholders or tenants, change their front doors in order to personalise their property. These replacement doors may not have the appropriate fire resistance rating required.

Responsible persons must ensure that signage is clearly visible to residents and users of the building and that means of escape are available to all persons including those with disabilities.

What is a higher-risk building under the Building Safety Act 2022?

Exact wording in the Act states

> 'higher-risk building' means a building in England that—
>
> (*a*) is at least 18 m in height or has at least seven storeys, and
> (*b*) contains at least 2 residential units.
>
> The Secretary of State may by regulations make provision supplementing this section.

The definition in the Act of residential unit is

> 'residential unit' means—
>
> (*a*) a dwelling, or
> (*b*) any other unit of living accommodation.

Therefore, in addition to residential buildings, care homes and hospitals can also be considered higher-risk buildings if they are higher than 18 m or have at least seven storeys, and so would fall under the new regulatory regime.

The Secretary of State has issued new regulations – The Higher-Risk Buildings (Descriptions and Supplementary Provisions) Regulations 2023 – that list exclusions to the general definition of higher-risk building listed above. The regulations state that buildings are not higher-risk buildings if they comprise entirely of

- a care home
- a hospital
- a secure residential institution
- a temporary leisure establishment or
- military barracks

The Higher-Risk Buildings (Descriptions and Supplementary Provisions) Regulations 2023 came in to force on 6 April 2023. These regulations determine which buildings will be subject to the new more stringent safety regime established by the Building Safety Act 2022 (the 2022 Act). They do so by supplementing and completing the definitions of higher-risk building at section 120D of the Building Act 1984 (the 1984 Act) and section 65 of the 2022 Act. The 1984 Act makes provision about the design of, and building work carried out in respect of higher-risk buildings. The 2022 Act makes provision about safety management in occupied higher-risk buildings.

What is a 'relevant' building under the 2022 Act, and how does fire safety apply?

Part 5 of the 2022 Act introduces a number of remedies to require landlords and associated persons to undertake and pay for remediation works for defects in 'relevant buildings'.

Relevant buildings are defined in section 117 as self-contained buildings (or self-contained parts of the building) containing at least two dwellings being at least 11 m high or having at least five storeys.

Relevant defects are defined in section 120 to include anything giving rise to a risk to the safety of people in or about the building due to the spread of fire or the collapse of the building or any part of it, and will also include anything done or not done during the 30-year period preceding the commencement of this section of the Act.

What are the recommended methods for carrying out a fire risk assessment in high-rise residential buildings?

As of May 2023, definitive government guidance on completing fire risk assessments in high-rise residential buildings has not been issued and the document 'Fire safety in purpose-built blocks of flats', issued by the Local Government Association (LGA) in 2011 and updated in 2020, is still considered applicable guidance.

Where a detailed assessment of external walls of existing multi-storey, multi-occupied residential building is deemed necessary, it should now be carried out in accordance with the more comprehensive and holistic guidance included in Publicly Available Specification (PAS) 9980.

Following the Grenfell Tower Inquiry, PAS 9980 has been developed by the British Standards Institution, drawing on expert advice from professionals across industry and following a rigorous development process, including a public consultation.

PAS 9980 provides new guidance on how to assess the risk of fire by way of an external wall of an existing multi-storey, multi-occupied residential building. PAS 9980 sets out steps that can be taken to identify and assess risk factors as well as mitigation steps that might improve the risk rating of a building by way of a holistic and fact-based assessment of a building's construction.

Where it is determined that a detailed assessment of an external wall is required, PAS 9980 should now be used for these assessments. It does not contain 'off the peg' solutions to specific wall types and materials, but is intended to enable a consistent approach to evaluating the fire risk when considering the external walls of buildings.

For details of the four types of fire risk assessment for high-rise residential buildings, see Chapter 4.

There are four types of fire risk assessment for high-rise residential buildings depending on the complexity of the building and knowledge of the building materials. Since the Grenfell fire, it is imperative that owners, landlords, property managers and so on consider the risks of fire spread from external wall cladding and insufficient fire doors to individual dwellings. There is also a need to assess any insulation used within the external framework of buildings as it is now known that fire can spread by way of the insulation and jump from storey to storey.

Type 1 – Common parts only (non-destructive)

A Type 1 fire risk assessment is the basic fire risk assessment required for the purpose of satisfying the RRO.

The inspection of the building is non-destructive. But, as well as considering the arrangements for means of escape, the fire risk assessment includes examination of at least a sample of flat entrance doors. It also considers, so far as reasonably practicable, the separating construction between the flats and the common parts without any opening up of construction.

However, in this type of fire risk assessment, entry to flats beyond the area of the flat entrance door, is not involved.

Where there are demountable false ceilings in the common parts, it may be appropriate to lift a sample of readily accessible false ceiling tiles. In addition, it will normally be appropriate to open a sample of service risers, provided access is practicable at the time of inspection.

Unless there is reason to expect serious deficiencies in structural fire protection – such as inadequate compartmentation, or poor fire stopping – a Type 1 inspection will normally be sufficient for most blocks of purpose-built flats. Where doubt exists in relation to these matters, the action plan of a Type 1 fire risk assessment may recommend that one of the other types of fire risk assessment be carried out or that further investigation be carried out by specialists.

Type 2 – Common parts only (destructive)

The scope and objectives of a Type 2 fire risk assessment are generally similar to those of a Type 1 fire risk assessment, except that there is a degree of destructive inspection carried out on a sampling basis. This will usually necessitate the presence of a contractor for the purpose of opening up construction and making good after the inspection.

In order to check the integrity of separating construction, the areas in which destructive inspection is carried out might sometimes include a sample of flats. However, because of the nature of the work, this can often only be carried out in vacant flats.

A Type 2 fire risk assessment is usually a one-off exercise, which is carried out only if there is good reason to suspect serious structural deficiencies that could lead to spread of fire beyond the flat of fire origin. The age of the block alone is not generally sufficient to warrant a Type 2 inspection. The need for a Type 2 fire risk assessment may sometimes be identified in a Type 1 fire risk assessment but should not simply be recommended as a matter of course.

Type 3 – Common parts and flats (non-destructive)

A Type 3 fire risk assessment includes the work involved in a Type 1 fire risk assessment, but goes beyond the scope of the RRO (though not the scope of the Housing Act). This risk assessment considers the arrangements for means of escape and fire detection (for example, smoke alarms) within at least a sample of the flats. Within the flats, the inspection is non-destructive, but the fire resistance of doors to rooms is considered.

Measures to prevent fire are not considered unless (for example, in the case of maintenance of the electrical and heating installations) the measures are within the control of, for example, the landlord.

A Type 3 fire risk assessment may sometimes be appropriate for rented flats if there is reason to suspect serious risk to residents in the event of a fire in their flats. This might be, for example, because of the age of the block or reason for suspicion widespread, unauthorised material alterations. This type of fire risk assessment will not be possible in the case of long leasehold flats, as there is normally no right of access for freeholders.

Type 4 – Common parts and flats (destructive)

A Type 4 fire risk assessment has the same scope of work as a Type 3 fire risk assessment, except that there is a degree of destructive inspection in both the common parts and the flats carried out on a sampling basis. This will usually necessitate the presence of a contractor for the purpose of opening up construction and making good after the inspection.

However, the nature of the work is such that, often, destructive inspection within flats can only be carried out in those that are vacant. This is the most comprehensive fire risk assessment but will only be appropriate in limited circumstances, such as when a new landlord takes over a block of flats in which the history of works carried out is unknown and there is reason to suspect serious risk to residents from both a fire in their own flats and a fire in neighbours' flats.

(NB: the above four types of risk assessment are reproduced from the LGA Guide referenced above.)

What needs to be assessed regarding fire doors to individual dwellings?

A fire door ensures that should a fire break out, it can be contained in a 'compartment'. This keeps the fire and smoke trapped for a defined period, allowing time for people to get out and make the fire easier to tackle.

It will not fulfil this function if damaged or propped open.

Signs that might indicate a fire door include things like blue 'Fire door' or 'Keep closed' signs, door closers and intumescent or smoke seals around the edge of the door or the frame.

In blocks of flats, the external door to a flat invariably should be a fire door to protect the common areas from spread of flame and smoke. Other locations will depend on the risk assessment and fire plan of the buildings. Internal doors could well be fire doors, depending on the size of the apartment and distance from the flat entrance door.

All fire doors are fire rated. Some are FD30 (providing 30 min protection), FD60 (60 min protection) or higher. There is usually a certification mark (a label or plug) on top of the door if it is a fire door.

Fire doors should be checked regularly, and the more they're used the more frequently they should be checked. Anyone can spot a dodgy fire door. But if you have legal responsibility for fire safety, call in a professional.

Create a fire door maintenance checklist and schedule and check all doors in your building. Only ever replace damaged components with like-for-like. Check the fire certificate. A trained person should be responsible for this maintenance work.

Five Point Check for Fire Doors

1. Certification. Check that the fire door is certified. There will usually be a label or plug on the top or side of the door. If the door does not include a label or plug, there is no evidence of certification, and this will need to be reported to the owner of the building.

2. Gaps. The gaps around the top and sides of the door should be no more than 3 mm, plus or minus 1 mm; so no greater than 4 mm and no less than 2 mm. To check if doors meet this standard, use a £1 coin as a guide, as this is approximately 3 mm wide. In most instances the gap at the bottom of the door can be as much as 8 mm. If daylight is seen under the door, the gap is likely to be too big. Where gaps are too big, cold smoke and fire will likely travel through the cracks rendering the fire door non-compliant. Any gaps that exceed this should be reported to the responsible person.

3. Seals: The intumescent seals around the door are designed to expand when they come into contact with heat. This helps to stop the fire from spreading through the cracks and seals the door shut. If the door does not have seals, or if the seals are faulty, report it to the owner of the building or the responsible person.

4. Hinges: Check that all hinges on the door are firmly fixed and that no screws are missing. If a door or the working parts – such as hinges that are not correctly maintained – it may not be capable of offering the necessary protection in the event of a fire.

5. Closing: A fire door will only offer the necessary protection if it is closed correctly by a door closer or is locked shut, as would be a store cupboard or riser door. If the door has a door closer fitted, the door can be tested by opening it to about half-way and then letting go. If the door closes fully by itself without catching on the floor or frame, it is fine. If it does not close correctly, report it. A fire door is no use to anyone if it does not close properly or is wedged open.

What is cladding and why is it important for fire safety?

The aluminium composite external cladding on Grenfell Tower caught fire and fuelled the spread of fire up the external facade of the building.

The term cladding, in its most common usage, refers to the outer skin(s) applied to a high-rise building to increase thermal energy efficiency and/or to improve aesthetics, while not adversely affecting weather resistance. The cladding element is non-load-bearing, which means it is not structurally integral to the building itself (it does, however, play a structural role, transferring wind loads, snow loads and its own self-weight, for example, back to the structural framework).

Cladding can either be fitted to an existing building of traditional masonry construction or can be incorporated into the design of a new building.

What are the different types of cladding?

There are many different types of cladding available, ranging from the traditional-looking brickwork or rendered systems to more modern metallic rainscreen systems or curtain walls made from glass. Cladding panels can be made from a wide variety of materials such as wood, metal, brick or vinyl, and are often coupled with composite materials that can include aluminium, wood, cement and recycled polystyrene.

The most common types of cladding facades are

- aluminium composite material (ACM) facade panels, which are made up of two aluminium skins bonded to a non-aluminium core, often to provide thermal insulation. This is the type of facade panel that was used in the cladding on Grenfell Tower, which caught fire in West London in 2017
- high pressure laminate (HPL) facade panels, which are made by pressurising layers of wood or paper fibres into a resin and then bonding them together using heat. This is the type of facade panel that was used on the University of Bolton student accommodation block that caught fire in 2019.

It is often hard to tell what the material make-up of a cladding system is purely by looking at the facade panels – often, further investigative work will need to be carried out.

How is cladding dangerous?

In fire safety terms, cladding can be dangerous in a few ways. Firstly, there is the make-up of the cladding system itself.

If the cladding system in question contains combustible materials, such as expanded polystyrene (EPS) or polyurethane (PUR), then exposing these materials to a flame will result in fire spread on the outside of a building. This can occur through either a fire breaking out from one of the windows or doorways, through ventilation shafts or from a fire that has occurred on the outside.

Fire stopping measures are incorporated into nearly every cladding system; however, some are more effective than others. Many are made using mineral-based materials, which are non-combustible and highly effective at slowing the spread of fire around the outside of the building.

Rainscreen cladding systems require an open cavity to allow the insulation to 'breathe' behind the weather-resistant cladding panels. This is potentially dangerous in a fire, as it creates a chimney for the fire to spread vertically through the building. For this kind of cladding to be used, most cavity barriers on the market now incorporate a strip activated by the application of heat that seals the cavity and, when applied correctly, dramatically slows the vertical spread of fire. This is called an intumescent cavity barrier.

Is cladding on every building?

No, cladding is not on every building. However, it is becoming increasingly commonly used on both new build high-rises and older concrete buildings that have been renovated. That said, just because a building has cladding on it does not necessarily mean that it poses a threat to its occupants. There are many different materials used in cladding systems and many modern materials are perfectly safe from a fire spread perspective.

What do the Building Regulations require regarding cladding?

The Building Regulations were amended in 2018 to implement a ban on combustible building materials on the outside walls of buildings that are taller than 18 m and contain more than one dwelling.

This ban applies to residential housing tower blocks, hospitals, residential care homes and student accommodation blocks.

Fire risk assessments for high-rise tower blocks will also start to include the external walls of a building.

Should a fire test be carried out on all cladding?

That depends on what the cladding is made from. In most cases, records of the materials that have been used in the construction process are held by the building owners or construction contractor – these will give an indication of whether the building is at risk.

If this information is not available, a sample of the cladding system may need to be tested in order to determine the materials that it contains, and whether there is a potential fire risk.

Owners/managing agents of high-rise buildings are advised to consult PAS 9980:2022 Fire Risk Appraisal of External Wall Construction and Cladding of Existing Blocks of Flats. Code of practice, published in 2022. PAS 9980:2022 is a code of practice aimed specifically at the safety of existing blocks of flats. Its purpose is to provide a methodology and framework for conducting appraisals of the fire risk posed by combustible external wall construction and cladding.

What changes have the Fire Safety (England) Regulations 2022 introduced regarding external cladding?

As of January 2023, the responsible person shall provide information to the local fire and rescue service (FRS) about the design and materials of a high-rise building's external wall system and inform the FRS of any material changes to these walls. They will also be required to provide information in relation to the level of risk the design and materials of the external wall structure gives rise to and any mitigating steps they have taken.

A template information form is given at the end of this chapter.

Case study: Fire safety defects in high-rise block of flats

Lambeth Council has been issued an enforcement notice by the London Fire Brigade after firefighters found considerable fire safety defects at a 20-storey tower block.

The failings identified at the council-owned building in Kennington, South London, included, among other things, 'failure to provide and/or maintain adequate and clearly indicated emergency routes and exits that lead to a place of safety' and 'failure to take general fire precautions to ensure the safety of persons on the premises'.

The enforcement notice was issued under the RRO. The London Fire Brigade made the following requirements known, which will need to be complied within a six-month period

- implement the general fire precautions as identified by the fire risk assessment
- implement a regular programme to review the fire risk assessment
- implement an arrangement for the effective planning, organisation, control, monitoring and review of the preventive and protective measures
- ensure that adequate escape routes and exits leading to a place of safety are provided and that these are always maintained clearly and available for use when the premises is occupied
- ensure that adequate maintenance systems are in place to ensure that the premises and any facilities, equipment and devices for use by, or protection of, firefighters are maintained in an efficient state, in effective working order and in good repair.

KEY RESOURCES

Building Safety Act 2022
Fire Safety Act 2021
Fire Safety (England) Regulations 2022

Regulatory Reform (Fire Safety) Order 2005
Building Regulations Fire Safety: Approved Document B
BS 9991:2015 Fire safety in the design, management and use of residential buildings. Code of practice – under review 2023
PAS 9980:2022 Fire Risk Appraisal of External Wall Construction and Cladding of Existing Blocks of Flats. Code of practice
Gov.uk: Fire safety in purpose-built blocks of flats: 2011 but due to be updated and republished in 2023
Fire Safety (England) Regulations 2022: Factsheets – various. Gov.uk 2022

External wall information for fire and rescue service as required by Fire Safety (England) Regulations 2022

1. Building identification

1.1. Name, address and postcode of building
1.2. Name and contact details of responsible person

2. Timber construction

2.1. Are structural timber systems used in the construction of the external walls?	
	Yes
	No

3. Masonry construction

3.1. Are the external walls constructed from masonry materials?			
	Yes		
	No (go to section 4)		
3.2. Is there any form of cladding or finish present over the outer masonry layer?			
	Yes		
	No (to report additional wall systems go to section 4, otherwise go to section 5)		
3.3. Select external facing materials present over the outer masonry layer			
	Aluminium composite materials		Metal sheet panels
	Other metal composite materials		Render system
	Brick slips		Stone panels
	Glass		Tiling systems
	High pressure laminate (HPL)		Timber
	Other (please specify)		

3.4. Select materials used for insulation between external facing material and masonry layer			
	Mineral wool		Phenolic foam
	Glass wool		Polyisocyanurate (PIR) or polyurethane (PUR) foam
	Expanded polystyrene (EPS) or extruded polystyrene (XPS)		None
	Other (please specify)		

3.5. Are these walls likely to ignite and spread fire easily? Consideration should be given to the combustibility of the external facing material, combustibility of any insulation, and any defects with the design and construction methods (e.g. issues with cavity barriers).	
	Yes
	No (to report additional wall systems go to section 4, otherwise go to section 5)

3.6. Outline the reasons why the walls are likely to ignite and spread fire easily

3.7. Identify the location of the walls, or sections thereof, which are likely to ignite or spread fire easily In some instances, the risk of external fire spread will be uniform across a building, in others, the risk will be limited to areas where specific materials have been used (e.g. certain floors or elevations).

If there are additional non-masonry external wall systems to report, then continue to section 4, otherwise go to section 5.

4. Alternative external wall systems

Although only presented once below, the questions in section 4 should be answered once for each different external wall system incorporated into the building design – that is, section 4 may need to be repeated. This is to allow clear differentiation between multiple external wall systems and their associated risk.

4.1. Select the external facing material			
	Aluminium composite materials		Metal sheet panels
	Other metal composite materials		Render system
	Brick slips		Stone panels
	Glass		Tiling systems
	High pressure laminate (HPL)		Timber
	Other (please specify)		
4.2. Select material used for insulation			
	Mineral wool		Phenolic foam
	Glass wool		Polyisocyanurate (PIR) or polyurethane (PUR) foam
	Expanded polystyrene (EPS) or extruded polystyrene (XPS)		None
	Other (please specify)		
4.3. Is this external wall system likely to ignite and spread fire easily? Consideration should be given to the combustibility of the external facing material, combustibility of any insulation and any defects with the design and construction methods (e.g. issues with cavity barriers).			
	Yes		
	No		
4.4. If yes, outline the reasons why the walls are likely to ignite and spread fire easily.			
4.5. Outline where on the building this external wall system has been used and, where necessary, how it can be distinguished from the other external wall systems that form part the design?			

5. Wall attachments and features

5.1. Does the building include any of the following attachments? Select all that apply			
	Balconies		Photovoltaic panels
	Green walls		Solar shading devices
5.2. Where the attachments selected above are likely to contribute to external fire spread, provide further information below			

6. Risk and mitigation

6.1. Following the building's fire risk assessment, was a further fire risk appraisal of the external walls required?	
	Yes, a further fire risk appraisal of the external walls has been completed
	Yes, a further fire risk appraisal of the external walls is required but not yet completed
	No, a further fire risk appraisal of the external walls was not required
6.2. What is the overall level of risk of fire spread due to the design and construction of the external walls?	
	Low risk
	Medium risk
	High risk
	The overall level of risk of the external wall has not been determined
6.3. What actions have been taken to mitigate the risk relating to the external wall?	
	Change to simultaneous evacuation strategy
	Remediation works to external wall
	Installation of sprinklers
	Removal of gas supply
	No additional measures are necessary

7. Person completing report

7.1. Name and contact details of person completing report

(NB: from government guidance document: Understand your duties as a responsible person under the Fire Safety (England) Regulations 2022.)

Perry P
ISBN 978-0-7277-6647-2
https://doi.org/10.1680/fsbqa.66472.169

Chapter 11
Fire safety on construction sites

What fire safety legislation applies to construction sites?

The following legislation may apply to fire safety on construction sites

- Regulatory Reform (Fire Safety) Order 2005 (RRO)
- Construction (Design and Management) Regulations 2015
- Management of Health and Safety at Work Regulations 1999
- Dangerous Substances and Explosive Atmospheres Regulations 2002
- Building Safety Act 2022.

Construction sites are places of work and therefore the duties placed on the responsible person/employer to maintain their premises in a safe condition apply to sites.

Construction sites carry a high risk of fire as the nature of the work increases the hazards and risks that contribute to a fire starting.

What are the general requirements for fire safety on construction sites?

Planning for fire safety must include making available adequate resources in terms of time, materials and money for the provision and maintenance of suitable and sufficient general fire precautions for the duration of the project.

A management ethos should be adopted such that suitable procedures and standards are laid down and adopted by all parties concerned in the project regarding the prevention of fires.

The fire prevention and protection measures to be adopted must be set out in the fire risk assessments undertaken in compliance with the RRO.

The principal contractor shall require subcontractors to observe their duties relevant to national fire safety legislation to ensure that the fire risk and potential for damage have been properly assessed and are kept to a minimum during construction.

The fire risk assessments in respect of all construction sites must be reviewed periodically in compliance with the legislation. This is important and necessary due to the rapidly changing nature of the hazards on all construction sites.

Fire safety must be recognised as an integral part of the management strategy for every construction site. In each case, therefore, a designated person must be responsible for the fire safety management system and inspections.

Because of the increased risk to personnel on high-rise construction sites and other locations where the means of escape may be compromised in the event of a fire, these sites should be identified in the fire risk assessment and in these instances the individual who is responsible for the fire safety management system and inspections should be permanently based on the site.

What are the principles of prevention in respect of fire safety?

The RRO requires all responsible persons to undertake fire risk assessments of their premises, operations and any other business activity that could expose persons to risks to their safety.

If the risk assessment indicates that preventative and protective measures

are required, then those measures must be in accordance with the requirements of Schedule 1, Part 3 of the RRO.

The principles of prevention are

- avoiding risks
- evaluating the risks that cannot be avoided
- combating risks at source
- adapting to technical progress
- replacing the dangerous with the non-dangerous or less dangerous.
- developing a coherent overall protection policy that covers technology
- organisation of work and the influence of factors relating to the working
- environment
- giving collective protective measures priority over individual protective measures
- giving appropriate instructions to employees.

What is the site fire safety plan?

In compliance with fire safety legislation the person responsible for the fire safety management system and inspections must ensure that a fire risk assessment is undertaken and is reviewed and updated regularly as construction proceeds. This document forms a key element of the site fire safety plan.

The plan must detail, as a minimum

- the organisation and responsibilities for fire safety
- the arrangements for recording fire safety training given to all site operatives
- general site precautions, fire detection and alarm systems, temporary emergency lighting and portable fire extinguishers

- the locations of designated smoking areas, where they are provided in compliance with no smoking legislation
- the requirements for a hot work permit regime
- the location, construction and maintenance of temporary buildings and temporary accommodation
- an effective evacuation plan and procedures for calling the fire brigade
- fire brigade access, facilities and coordination
- the instructions given to those on site of the required actions in case of fire
- security measures to minimise the risk of arson
- the regime for the storage and control of waste materials; the regime for the storage and control of flammable liquids and compressed gases.

What is the role of the person responsible for fire safety on the site?

The person responsible for the fire safety management system and inspections on a construction site must

- ensure that all procedures, precautionary measures and safety standards as laid down in the site fire safety plan are clearly understood and complied with by everyone on the site
- ensure that a system for the issuing of hot work permits is established and monitored
- conduct weekly inspections of escape routes, fire safety signage and temporary emergency lighting (where applicable)
- carry out weekly tests of the site fire detection and alarm devices installed on site
- carry out weekly checks of firefighting equipment, fire brigade access and firefighting facilities
- carry out weekly checks of the routing of temporary electrical cables, the housekeeping on site and the storage of combustible waste materials
- conduct periodic fire drills to ensure that everyone on site is aware of the procedures and reacts appropriately. This includes rehearsing the procedure for alerting the fire brigade
- liaise with the local fire brigade and invite them to undertake site inspections and familiarisation tours where appropriate
- liaise with site security personnel where they are employed
- ensure that a proper maintenance regime for fire protection equipment is instituted, including the keeping of a written record of all checks, inspections and tests
- maintain a written record of training of site operatives and of all fire patrols and fire drill procedures
- where appropriate, appoint enough fire marshals, who should be properly trained to assist in the evacuation of the site and take first aid and firefighting measures where it is safe to do so
- during an emergency, execute those duties required for the safe evacuation of everyone on site, ensuring that all staff and visitors report to the assembly points
- take action to promote a fire safe working environment at all times.

What steps should be taken for liaising with the emergency services?

Liaison should be established with the local fire brigade and an initial site plan should be provided. Thereafter, updated site plans should be available at the entrances to the site for fire brigade use, detailing the following

- fire brigade access points to the site, firefighting shafts, fire lifts and temporary hoist facilities
- dedicated emergency escape routes and staircases
- sprinkler installations and the location(s) of their stop valves
- floor loading limitations
- positions of hydrants on or near the site, dry riser inlets and wet risers
- the location of temporary buildings and temporary accommodation
- the location of hazardous items including (but not restricted to) flammable liquids, gas cylinders, gas mains, electrical risers and temporary holes in floor slabs.

Where work on the site may have an impact on traffic movements in the vicinity, liaison should be established with the local police.

The location of suitable medical facilities should be recorded on site.

Where emergency rescues are not undertaken by the local fire brigade, details and locations of appropriate rescue services should be kept on site.

What consideration needs to be given to water supplies on the site?

In the case of large projects, or those where structures are being constructed predominantly of combustible materials (such as timber), the fire brigade should be informed and provisions for water supplies agreed before work commences on site.

- Adequate water supplies for firefighting must be available.
- Rising and temporary mains must be provided where planned.
- Water supplies should be tested periodically.
- It may be necessary to move the fire brigade inlet point to rising mains as work progresses.

In the case of large projects and those where the construction is predominantly of combustible materials, following the agreement for water supplies with the fire and rescue service, on-site water flow should be tested and recorded before work commences and thereafter every three months, at which time all valves should be exercised.

Should water supplies fall short of the pre-determined requirements at any time on large projects and those where the construction is predominantly of combustible materials, no hot work shall be undertaken until adequate flows have been restored.

All hydrants must be kept clear of obstructions and be suitably marked.

What should be considered regarding fire brigade access?

Fire brigade access routes to the site and buildings must be maintained clear and unobstructed at all times.

Significant changes to the access to the site should be discussed with the fire brigade before being implemented.

What emergency procedures are required for the site?

On all sites, a means of giving warning of fire must be established. Manual devices may be utilised provided that

- they are clearly audible above background noises in all areas
- they are distinctly different from any other warning devices on site and can be readily identified as being a fire alarm
- there are multiple provisions of these devices to ensure that they can be accessed at all times.

Written emergency procedures must be displayed in prominent locations and given to all employees on site.

Nominated personnel, such as the security guards, must be briefed to open gates or barriers and provide ready access to the site for the fire brigade in the event of an emergency or their other visits to the site.

Clear signs must be provided and maintained in prominent positions indicating the locations of fire brigade access routes, escape routes, positions of dry riser inlets and the fire extinguishers provided for use by trained staff. Signs should be reviewed regularly and replaced or repositioned as necessary.

In the event of a fire, contractors should determine that all of their personnel on site have been accounted for and pass this information to site security staff at the earliest opportunity.

Fire protection needs to be considered on the site. What should I do?

The contractor should ensure that the project is designed and planned in sequence to achieve the installation and operation of the following as early as possible and economically reasonable in all cases where they are to be provided

- permanent fire escape stairs, including compartment walls
- fire compartments within the building under construction, including the installation of fire doors, and the provision of temporary fire stopping to prevent the passage of heat and smoke
- fire protective materials to structural steelwork

- planned firefighting shafts
- lightning conductors
- automatic fire detection systems
- automatic sprinkler and other fixed firefighting installations
- automatic fire detection and extinguishing systems, where these are to be installed to protect large or costly items of equipment or plant.

To protect distribution panels and items of electrical equipment, appropriate extinguishers (such as those containing carbon dioxide) must be provided close to the equipment concerned.

At the end of each working day or shift a fire check must be undertaken, particularly in areas where hot work has been undertaken. Where 24 h security is provided, fire checks should be undertaken throughout the night, during holiday periods and at weekends.

Costly equipment such as large items of plant should not be installed until the fire protection systems have been commissioned.

When finished surfaces, fittings or expensive items of plant and machinery are to be temporarily protected during construction or refurbishment then care should be taken to select a protective covering material that will not contribute to the fire load or the potential for fire growth and spread.

When flexible materials are used to clad scaffolding, these materials should not contribute to the fire load or the potential for fire growth and spread.

Occupancy of any part of a building site by building owners or tenants should not be permitted until all fire protection measures (especially all fire stopping in relevant compartment walls and ceilings) and installations are complete and, where appropriate, have been commissioned.

Construction workers must not live on site. Caravans or similar purpose-built accommodation should be separated from the building under construction or refurbishment and be enclosed by a palisade, fence or hoarding such that there is no interconnecting route between the two areas.

What portable fire extinguishers are recommended for a construction site?

Personnel must be sufficiently instructed to be able to use the portable firefighting equipment provided on site.

There should be an adequate provision of appropriate portable fire extinguishers, approved and certificated by an independent third-party certification body, in accordance with national standards.

Where structures are being constructed predominantly of combustible materials, additional portable fire extinguishers must be provided in accordance with a specific fire risk assessment. This assessment should be recorded as part of the site fire safety plan.

Extinguishers must be located in conspicuous positions near exits on each floor. In the open they should be suitably protected from the environment and prominently signed with a pictogram.

All portable firefighting equipment must be serviced annually by a qualified person in accordance with national standards and the maintenance service date recorded, including marking on the appliances.

As work progresses, the adequacy of portable firefighting equipment must be reviewed as part of the periodic review of the fire risk assessment for the site.

All firefighting equipment that is not designed to come into use automatically must be easily accessible.

Wherever reasonably practicable, 'ride-on' mechanically propelled site plant should carry an appropriate fire extinguisher.

Does the site agent need to consider site security against arson when preparing the fire safety plan?

Arson protection should be a feature of the site fire safety plan and must be addressed as part of the fire risk assessment for the site.

Buildings must be suitably protected against theft and deliberate fire raising in accordance with the fire risk assessment.

The most effective method of deterring trespassers, as well as helping to prevent malicious fire, is to ensure, as far as reasonably practicable, that the site is secured against unauthorised entry. This may be achieved by erecting a suitable hoarding around the perimeter of the site or securing all access points such as windows and doors on refurbishment sites.

Flammable liquid stores, liquefied petroleum gas cylinder storage and combustible material stores must be fenced or otherwise suitably protected.

Illumination of the site is an effective deterrent to unauthorised access and is recommended.

The installation of CCTV cameras should be seriously considered on sites where arson, vandalism or theft may occur.

The recruitment of security personnel should be considered on all sites, especially for employment on site outside normal working hours.

A permanent security presence must be provided on sites where the construction is predominantly of combustible materials.

The installation of intruder alarms in temporary buildings and temporary accommodation is strongly encouraged.

What should be considered for temporary buildings and temporary accommodation?

The site fire safety plan must include a suitable and sufficient fire risk assessment for all temporary buildings and temporary accommodation. The assessment should be reviewed periodically.

Temporary buildings should be separated from the building under construction or refurbishment and other permanent buildings to provide a fire break which, where possible, should be at least 10 m, and 20 m in the case of sites where the structure under construction is predominantly of combustible materials. Similarly, rows of temporary buildings should be separated in a similar manner to provide a reasonable fire break.

Glazing must be fixed shut and provide a degree of fire resistance as determined by a risk assessment where the glass is within 20 m of another building or structure.

Where floors of temporary buildings are raised above ground level, the space beneath must be enclosed to prevent accumulation of rubbish, while still allowing under-floor ventilation. No combustible materials should be stored under a temporary building.

Temporary buildings must not be constructed within a building or structure under construction or refurbishment.

Automatic fire detection systems must be installed in temporary accommodation and in temporary buildings when these are

- within 6 m of a building under construction or refurbishment
- used for the storage of flammable liquids or compressed gases
- used for the drying of clothes.

Temporary buildings or temporary accommodation where cooking (including the use of toasters) is undertaken should have automatic fire detection installed.

Heaters for use in temporary buildings and temporary accommodation must

- have enclosed elements
- be fixed in position, preferably above floor level
- be fitted with securely fixed metal guards
- be thermostatically controlled
- be maintained in a sound condition.

Carelessly drying of clothes causes fires. Coat stands and drying racks must be firmly positioned at a safe distance from heaters.

All heaters and cooking appliances must be properly installed with adequate ventilation provided. Microwave ovens are preferable to proprietary electric or gas cookers.

Consideration should be given to the installation of automatic sprinkler systems in temporary buildings on sites where construction is predominantly of combustible materials.

Temporary accommodation should not be provided in structures composed predominantly of combustible materials.

Temporary buildings and temporary accommodation should not contain more than the minimum of combustible furniture and fittings.

What is good practice regarding site storage of flammable liquids and gas cylinders?

Flammable liquids, compressed gases and/or LPG must not be stored together.

Containers of flammable liquids, gas cylinders and LPG cylinders should preferably be stored in open compounds that are securely fenced, shaded from the sun and remote from pits, drains and low-lying areas.

Appropriately worded warning signs – such as ‘HIGHLY FLAMMABLE LIQUIDS’, ‘NO SMOKING’ and ‘NO NAKED LIGHTS’ – must be boldly displayed at the entrances to flammable liquid and gas cylinder stores.

Any electrical fittings – for example, lights and switches – within these stores must be suitable for an environment where a flammable or explosive atmosphere may be present and be selected and installed by competent persons.

Adequate numbers of extinguishers appropriate to the hazard should be sited at the entrances to storage areas.

Flammable liquids

Stores of liquid fuels must be situated on an impervious base and surrounded by a bund sufficient to contain the maximum contents of the largest drum stored plus 10 per cent. The bund must not be allowed to accumulate water or waste material.

Where it is necessary to store flammable liquids inside a building under construction or renovation, the quantity so stored must be the minimum necessary and no more than a day’s supply. The containers must be kept in a store, cupboard or bin that is of fire-resistant construction.

Ideally, storage areas should be sited at least 10 m from permanent and temporary buildings and 20 m from structures fabricated predominantly from combustible materials. Containers must not be stored within 4 m of any building or boundary fence unless the boundary is a wall with at least 30 min fire resistance and at least 2 m high. In the latter case, containers and drums should be at least 1 m below the top of the wall.

Products that could add to the intensity of a fire, such as acetylene or oxygen, or to the toxic hazard in the event of fire, such as chlorine, must not be stored in the same compound as flammable liquids and LPG.

Gas cylinders

The floors of LPG and other cylinder stores should be paved or compacted level with a suitable hard standing provided for the delivery and dispatch of cylinders. The area must be kept clear of all combustible materials, weeds and rubbish.

The provision of automatic flammable gas detection equipment should be considered for enclosed storage locations.

All permanent LPG supplies and their connections to LPG-fuelled appliances must be installed by a competent gas fitter.

Gas supply to appliances should be by fixed piping or armoured flexible tubing.

Gas cylinders should be located outside buildings and be secured and protected from unauthorised interference.

Gas appliances must be fitted with control taps.

Acetylene

Acetylene is a serious hazard on construction sites and elsewhere. It is a flammable gas that at elevated temperatures and pressures, or following impact of the cylinder, becomes unstable and liable to spontaneous combustion. As a result, acetylene in cylinders, once suspected to be unstable, constitutes a serious fire hazard.

In these circumstances, fire brigade safe working practices include the establishment of a hazard zone of up to 200 m around the incident and leaving the cylinders involved undisturbed for a period of up to 24 h or more prior to removal. All activities in the designated hazard zone have to cease and the area must be evacuated, with significant implications for construction operations and other businesses operating in the area.

The use of acetylene on construction sites should therefore be eliminated wherever possible and alternative methods of cutting and welding be adopted. Where the use of the gas is unavoidable, its presence must be minimised.

Where the use of acetylene is unavoidable, spare cylinders must not be kept on site.

Acetylene cylinders should be removed from the workplace and returned to the storage area as soon as the period of work has been completed. The cylinders should be removed from the site as soon as their use is complete.

Gas cylinders should always be adequately supported, preferably by mounting on purpose-built trolleys.

Equipment, hoses and flashback arrestors used with oxyacetylene and similar equipment should be in good condition, set up in accordance with the manufacturer's instructions and be subject to a visual inspection before each period of use.

Gas welding and cutting procedures should only be carried out by trained personnel.

The use of acetylene cutting and welding equipment must be subject to a hot work permit once fitting out work has commenced on site and in all buildings that are being refurbished.

Case study: Groundworks contractor fined for petrol fire (2022)

A Kent groundwork contractor was fined after a worker sustained serious burns following petrol being thrown on a bonfire during a groundworks project being undertaken in 2020.

A 26-year-old groundworker employed by a surfacing company was working on a new residential development in Kent, when a co-worker used petrol on a bonfire. The groundworker was unaware of this and after he was instructed to light the bonfire, it engulfed him in flames as the petrol vapour ignited. The worker suffered serious burns and underwent two skin graft operations to his left hand, left arm, left side of torso and both his legs.

The Health and Safety Executive found the company had failed to appropriately supervise their operatives and failed to provide them with the appropriate information and instruction, so far as is reasonably practicable, to ensure work was carried out without risks to health or safety.

At the local Magistrates' Court, the company pleaded guilty to breaching Regulations 15 (8) of the Construction (Design and Management) Regulations. They were fined £10 000 and ordered to pay costs of £7333.42.

Speaking after the hearing, the HSE Inspector said

> The operative's injuries are life changing and could have easily been fatal.
>
> This serious incident and devastation should have been avoided if those in control of the work provided the appropriate supervision, information and instructions to their workers.

What precautions are needed in relation to hot works on site?

Alternative methods to hot work should be adopted wherever possible. Where hot work cannot be avoided, the guidance set out in HSG168 Fire safety in construction and CFPA Europe guideline 12 should be followed.

Hot work should only be undertaken by suitably trained staff.

When there is no alternative to hot work, the hot work should be undertaken, where possible, in a dedicated area away from the main area of construction work or storage of materials.

A 'permit-to-work' system must be adopted where hot work is being undertaken unless there is no risk of damage to any surrounding property.

Advice is also given in various guides concerning the use of tar boilers. Tar boilers and similar equipment should be placed at ground level wherever possible. Only if a risk assessment shows that overall it is a greater hazard to have the boiler on the ground may it be placed in another location convenient for the works.

What precautions are needed in relation to electrical supplies and equipment?

Electrical supply installations, both temporary and permanent, must be installed in accordance with national standards.

All electrical installations must be undertaken by a competent electrician.

A competent electrician should inspect all installations, especially those of a temporary nature, regularly and test them at intervals not greater than every three months and following all alterations. The results of tests and inspections should be recorded.

Electric cabling, especially temporary installations, should be protected against damage from construction site activities in the vicinity.

Portable electric equipment used on site shall carry durable labels that state it is subject to periodic inspection and testing.

Where portable or temporary lights are required, these should be located well away from combustible materials.

Where possible, main switches, other than those controlling security and automatic fire detection systems, should be turned off when work ceases and all equipment be unplugged when not in use.

Are there any fire risks with plant and vehicles?

Stationary plant powered by internal combustion engines, such as compressors and generators, should, wherever practicable, be positioned in the open air or in a well-ventilated non-combustible enclosure. They must be sited so that exhaust pipes and exhaust gases are kept clear of combustible materials.

Air intakes must be situated so that the air is cool, uncontaminated and free from flammable gases or vapours.

Fuel tanks must not be filled while engines are running.

Compressors should be housed singly away from other plant and in a separate enclosure. Plant and equipment must be protected against accidental impact.

Where appropriate, sand trays should be provided to absorb drips of fuel or lubricant. The trays should be changed at regular intervals.

As a general rule, the long-term parking of vehicles should not be permitted within 10 m of the building under construction and, if possible, a separate car park should be available for workers' vehicles.

Under no circumstances should long-term parking be permitted within the building under construction without a suitable and sufficient fire risk assessment being undertaken.

When equipment and materials are being unloaded from or reloaded onto contractors' and subcontractors' vehicles, such vehicles may be permitted to park on site within 10 m of the building, but for no longer than the duration of unloading or reloading.

What precautions are necessary regarding stored and waste materials?

Where it is reasonably practicable to do so, combustible materials should be stored outside the building under construction or undergoing refurbishment and should not be so close to it that fire is able to spread from the materials to the building. Wherever possible, stored combustible materials should be at least 10 m away from a partially complete structure, and 20 m in the case of a predominantly combustible structure.

Waste materials must not be stored within a building under construction or refurbishment.

Where combustible construction materials are stored inside the building, the area used for storage should

- have controlled access
- not be in an area where hot work is being carried out
- either be within the area covered by the site fire detection system or be included on the route of regular fire checks
- have firefighting equipment located close by.

Protective covering materials should be available to cover combustible construction materials during storage. As indicated above, protective covering materials should not contribute to the fire load or the potential for fire growth and spread.

Good housekeeping is essential on all sites as waste material, including all non-essential combustible wrapping and packaging, if allowed to accumulate, provides an excellent starting point for fire. Therefore, the introduction of combustible waste should be minimised and all combustible waste, packing materials, wood, shavings and oily rags must be regularly removed from the buildings and disposed of at the earliest opportunity.

Where rubbish chutes are to be used, they should be constructed outside the building and be of fire-resisting construction. Combustible materials should not be allowed to accumulate in or near the chute.

Where practicable, all collected waste materials awaiting disposal must be kept in an area at least 10 m away from the structure upon which work is being undertaken, temporary accommodation, smoking shelters, stores and equipment. On a site where a predominantly combustible structure is being erected, waste materials should be at least 20 m from the structure.

Separate metal bins, with close-fitting metal lids, must be provided for flammable materials – for example, oily rags.

All dry vegetation must be cleared regularly from within 20 m of all buildings or structures. Rubbish must not be burned on site.

Can smoking be allowed on a construction site?

A 'no smoking' policy must be established on the site except for designated outdoor areas where smoking will be allowed in accordance with current legislation.

Where a smoking shelter is provided it must comply with the UK smoke-free legislation and be

- subject to a specific fire risk assessment
- constructed of non-combustible material
- where practicable, sited at least 10 m away from any building or structure (20 m on a site where a predominantly combustible structure is being erected)
- provided with suitable metal ashtrays and a separate metal waste bin with a fitted metal lid
- provided with a suitable fire extinguisher.

The immediate area around the shelter and the shelter itself should be kept clear of combustible materials including windblown debris and vegetation.

Raised, slatted floors or decking should not be used, and concealed or semi-open spaces should be sealed to ensure combustible debris cannot accumulate beneath the shelter.

Combustible curtains, canopies and drapes must not be used to protect smokers from the elements.

In no circumstances should the shelter be sited near

- windows
- ventilation intakes or extracts
- entrances and exits from the premises
- hazardous materials
- waste storage containers – such as skips or bins
- beneath a canopy or low-slung eaves.

Areas where smoking is allowed but no shelter is provided must be free of combustible materials and be equipped with firefighting equipment, metal ash trays and a separate metal waste bin with a fitted metal lid.

No smoking must be permitted in outside areas where fire hazards exist. Such areas will include refuse and storage areas containing combustible materials, flammable liquids (including refuelling supplies), gas cylinders, foam plastics, fibreboard and timber. 'NO SMOKING' notices must be displayed prominently in these areas.

What are the fire risks with high-rise construction sites?

There are a number of sites where construction progresses at heights such that the time taken to escape from the upper levels to a place of safety away from the building in an emergency is excessive and rescue by the fire brigade using conventional ladders and hydraulic platforms may not be possible.

The problem may be compounded by inadequate water supplies to fight a fire at a high level and incomplete compartmentation of the building might lead to an inordinately rapid spread of the flames. In these circumstances additional provisions are required to ensure the safety of staff as well as minimising the extent and cost of the damage.

When work on a site reaches a height such that the 'high-rise construction site' level has been reached

- a wet riser should be provided fed by duplicate pumps
- fire doors with self-closers must be fitted to protect the escape stairs and, where applicable, access to the shaft in which the crane is situated
- the building should be horizontally fire compartmented at intervals not exceeding ten floors to prevent the spread of smoke and flames. This should be done at the earliest opportunity after construction of each of the relevant floors, using temporary fire-stopping materials having not less than 1 h fire resistance until the permanent fire stopping arrangements can be put in place.

What precautions are necessary on sites where construction is of predominantly combustible materials?

The building should be compartmented at the earliest stage possible. To accomplish this, fire resistant sheeting to the full height of the compartment walls, to form compartments at a maximum of 20 m centres horizontally, should be fixed in place during the erection of each storey.

The proximity of the incomplete structure to the site boundary and to surrounding buildings should be considered as part of the fire risk assessment.

Serious consideration should be given to mitigating fire damage by facing exposed timber construction and combustible insulation with non-combustible materials at the earliest opportunity. The use of such materials may be extended to protect windows and door openings not required as means of escape. This approach also provides significant security benefits.

Even when the perimeter of the site is secure, access to the building under construction should be denied when work is not in progress.

Case study: Nottingham University GSK building fire (2014)

The University of Nottingham's wooden-framed laboratory, designed to be carbon neutral, was wrecked by a blaze on 12 September 2014.

The Health and Safety Executive said it was looking at construction working practices on site but stressed this was routine in such cases.

The wooden-framed structure had been completed and was due to open the following year.

Sixty firefighters were needed to put out the blaze at the GlaxoSmithKline building.

Early assumptions could be made relatively quickly into the process because the building was still under construction. At the time of the fire, it was found that the building did not have a mains gas or mains electrical supply. As such, ignition related to the mains supplies from the utilities could be discounted.

What is of note is that there was a temporary power supply to the building for primary lighting, emergency lighting and for transformers to power electrical tools and charge mobile working platforms. While all of these items were portable appliance tested, at this stage they could not be discounted as a possible source of ignition.

As a result of the evidence gathering process and the discounting of possible ignition sources, electrical faults as a cause became the most probable cause. Although a fault with the mains utility supply had previously been discounted, it was established that there was a temporary electrical supply into not only the site, but also into the building under construction itself.

With the support of the principal contractor, the investigation secured plans of the route that the temporary cable installation took and the nature and type of electric construction equipment that was in each area of the building. By surveying the scene post fire, the remains of these items were identified, marked and photographed in situ.

Over the days that followed and through the cross-checking of the evidence and data available, it was finally determined that the most probable cause for the fire was electrical, and the origin or location of the fire was on the first floor.

At this stage of the construction, without fire doors or, in some areas, glazing, there were open voids between floors. While services were being installed and commissioned, this caused the building to be self-ventilating and once the fire had taken hold it passed through the building rapidly and with some ferocity.

What are the steps I need to take to complete a fire risk assessment for the construction site?

The process of a fire risk assessment on a construction site is the same as for any other building or work environment.

The risk of a fire may be greater on a construction site due to the accumulation of combustible materials and the working practices carried out on the site.

Start the fire risk assessment by adopting the five steps of risk assessment

- identify the hazards
- identify the people and the location of people at significant risk from a fire
- evaluate the risks
- record findings and actions taken
- keep assessment under review.

Identify sources of ignition

You can identify the potential ignition sources prior to and during the construction process by looking for possible sources of heat that could get hot enough to ignite material found on your site. These sources could include

- smokers' material – for example, cigarettes, matches and lighters
- naked flames – for example, gas- or liquid-fuelled open-flame equipment
- those deliberately introduced (arson)
- bonfires
- plant and equipment – for example, fuel and vehicle exhausts
- electrical – faulty or misused electrical equipment
- poor electrical installations – for example, overloads, heating from bunched cables and/or damaged cables
- hot processes/hot work – for example, welding by contractors
- light fittings and lighting equipment – for example, temporary lighting or halogen lamps too close to stored products
- electrical, gas or oil-fired heaters (fixed or portable), room heaters in temporary office accommodation or welfare cabins
- heat sources such as gas, electric, cooking equipment, microwaves
- friction-generated heat from mechanical equipment such as disc cutters
- static charge from mechanical equipment
- use of oxy-fuel equipment
- spontaneous ignition and self-heating – for example, oil-soaked rags and paint scrapings
- lightning and refracted sunlight.

Identify sources of fuel

Anything that burns is fuel for a fire. Many materials that can burn have to be used during construction work. Reducing the quantity of material on site reduces the chances of fire occurring and limits the extent of any fire that should start. Stocks of high fire hazard material should be managed to balance production needs with the need to reduce the risk of fire. Limit the material present at worksites to what is needed for half a day or a single shift and return unused material to the stores when the work is finished.

Where combustible or flammable materials must be used, select the least combustible alternatives.

Look for the things that will burn and are in enough quantity to provide fuel for a fire or cause it to spread to another fuel source. Some of the most common fuels found on site include

- components of the structure itself such as some composite panels and timber
- stored/in-use building products such as composite panels and timber
- rubbish
- flammable liquids such as paints and varnishes
- protective coverings
- scaffold sheeting
- volatile flammable substances such as paints and thinners
- fuel for portable equipment
- liquefied petroleum gas (LPG) – for example, bitumen boilers
- temporary site accommodation and similar areas
- acetylene
- packaging materials
- petrol disc cutters and other portable equipment
- fall-arrest bags.

Identify sources of oxygen

The main source of oxygen for a fire is in the air around us. On construction sites this will be natural airflow through doors, windows and other openings. Wind or the 'chimney effect' can also cause increased oxygen to feed the fire.

Additional sources of oxygen can sometimes be found in site processesor materials used or stored on site such as oxidising agents; they can provide a fire with additional oxygen and so help it burn. These chemicals should have identification on their container (and a COSHH/CHIP data sheet) by the manufacturer or supplier who can advise as to their safe use and storage.

Examples include

- oxygen used in welding processes
- oxidising agents.

Identify those at risk from fire

All operatives on the site will be at risk from fire as will contractors, delivery drivers, agency staff, casual workers, visiting clients and professional design teams.

Evaluate the consequences and risks of a fire on the site

What are the likely outcomes of a fire on site?

How will people be at risk from the fire – for example, affected by inhalation of smoke or toxic fumes, unable to evacuate, cut off from escape routes, unable to hear the alarm, unable to raise the alarm, lighting failures so unable to see to evacuate the building or suffering from a disability that prevents them taking action.

How likely is a fire? How quickly will it spread? How protected are the escape routes?

Record the findings of the assessment

Use a standard template fire risk assessment form to record all the findings from the assessment.

Identify the control measures necessary to minimise the risk of a fire on the site

Decide on what steps can be taken and by whom in order to reduce the likelihood of a fire starting on the site.

Devise a fire safety plan for the site.

Implement an evacuation plan.

Include fire safety and evacuation in induction training.

Place suitable fire extinguishers around the site.

Install a fixed temporary fire alarm system where possible and, at the very minimum, hand bells or klaxons in all areas.

Install temporary emergency lighting around the site as appropriate.

Implement regular fire safety checks across the site and maintain all fire escape routes.

Appoint fire wardens/marshals to conduct fire watches and to take control of any localised incident.

Manage hot works including welding activities.

Manage waste and materials storage.

Regularly inspect electrical equipment.

Control the use of flammable products.

Use only fire-retardant/flame-resistant temporary coverings and hoardings.

Consider the fire management plan for high-rise buildings and consider a phased evacuation of floors to prevent congestion on any escape staircases.

Establish a continued source of water so that firefighting can be carried out by the fire and rescue service.

Establish contact and liaison with the local fire and rescue service and advise them of any exceptional fire risks on site.

Review assessment

Regularly review the findings of your fire risk assessment and update it when any conditions change.

What extra precautions may be necessary for refurbishment works?

The incidents of fire on refurbishment sites are more frequent than on new build sites because there is often a greater amount of combustible materials on the site due to existing materials and furnishings.

The structural fire protection elements of the building may be compromised during refurbishment works, thus causing a fire to spread rapidly through the building.

Existing fire detection systems may be temporarily deactivated within the area being refurbished, thus causing the system to fail in other parts of the building – for example, smoke detectors may be isolated, alarms deactivated and smoke curtains overridden.

A comprehensive fire risk assessment will be needed for the area where refurbishment works are being undertaken and this will need to be carried out in conjunction with an assessment of the existing building and how it will be used and the likely fire risks from both areas.

The fire risk assessment for the building undergoing refurbishment will need to be amended to take in to account the hazards and risks from the refurbishment works.

Communication between all parties is key and a combined fire safety plan should be developed that considers both the effect of fire on the building occupants if caused on the construction site and the effect of a fire if started within the building on the construction operatives.

KEY RESOURCES

Regulatory Reform (Fire Safety) Order 2005
Construction (Design and Management) Regulations 2015
Management of Health and Safety at Work Regulations 1999
Fire Prevention on Construction Sites: The Joint Code of Practice on the Protection from Fire of Construction Sites and Buildings Undergoing Renovation (10th edition) 2023
Fire safety in construction: HSG168 (3rd edition) July 2022

Checklist

Is the quantity of flammable materials, liquids and gases kept to a minimum?

Are they properly stored?

Are flammable gas cylinders returned to a ventilated store at the end of the shift?

Are smoking and other ignition sources banned in areas other than designated smoking areas?

Are gases or flammable liquids stored or used on site?

Are gas cylinders, associated hoses and equipment properly maintained and in good condition?

When gas cylinders are not in use, are the valves fully closed?

Is flammable and combustible waste removed regularly and stored in suitable bins or skips?

Are suitable fire extinguishers provided?

Are hot works suitably controlled by a Permit to Work or Hot Works Permit?

Have fire wardens been appointed?

Is the fire risk assessment complete?

Is fire signage complete around the site?

Has the assembly or muster point been identified?

Are there emergency procedures – for example, for evacuating the site in case of fire?

Do people on site know what the procedures are?

Is there a means of raising the alarm and does it work?

Is there a way to contact the emergency services from site?

Are there enough suitable escape routes and are these kept clear?

Perry P
ISBN 978-0-7277-6647-2
https://doi.org/10.1680/fsbqa.66472.191

Chapter 12
Fire safety in special and other premises

There are many buildings, venues, industries and services which could require special considerations for fire safety. It is not practicable to cover every special premises/industry or service provider in this chapter and so only three special premises types have been chosen to demonstrate the additional considerations that need to go into fire safety planning and management, namely

- residential and care homes
- schools, colleges and universities
- entertainment premises.

A wealth of information is available on fire safety management for

- hospitals
- shopping centres
- heritage buildings
- industrial plants especially chemical and oil refineries
- oil rigs and the offshore industries
- coal mines, mining and quarrying
- hotels and tourist accommodation
- caravans and camping sites
- sports stadia and football grounds.

And of course, after the Grenfell Tower fire, more specific information is available for managing fire safety in low-, mid- and high-rise residential buildings.

What extra steps do I need to take in respect of fire safety in residential and care homes?

The basic principles of managing fire safety apply to residential and care homes.

In addition to the normal risks of a fire starting and taking hold in a building, the responsible person of the home must consider specialist health equipment.

Home oxygen therapy

Used by people with severe respiratory conditions, oxygen therapy provides air that contains more oxygen than normal. Specialist equipment (which can include cylinders and portable units) is used to pipe medical oxygen through either a nasal cannula or face mask to the patient.

The addition of concentrated oxygen into the room or surrounding environment will greatly increase the intensity of a fire should one start.

Smoking or the use of any naked flames (such as candles, fires and cooking) anywhere near oxygen is extremely dangerous and increases the risk of significant injury in the event of a fire.

Make sure oxygen is turned off when not in use.

Dynamic airflow pressure relieving mattresses and overlays

These items are used for the prevention and treatment of pressure sores and ulcers that can be caused by extended periods of immobility. The mattress is filled with air by a pump, which adjusts pressure according to the patient's needs.

When punctured by any heat source (e.g. a cigarette or match), the escaping air can cause a fire to spread rapidly. The emergency back-up battery may also continue to pump air, which can cause the fire to burn longer and with greater intensity.

Incontinence products

These are often supplied in large quantities to people who have continence problems. They contain plastics and other chemicals, in addition to paper or textile padding, and can provide additional fuel to a developing fire.

Emollients and skin creams

Emollients and skin creams are an important and effective treatment used to prevent or treat dry skin conditions like

- eczema
- bed sores
- ulcers
- psoriasis.

They come in a variety of forms: creams, lotions, ointments, gels or sprays. They can also include soap alternatives. They may be water-based, contain paraffin or natural oils. All cover the skin with a protective film to reduce water loss.

Emollient/skin cream residue on fabrics such as bedding, clothing and dressings can increase flammability, even from just one application. Emollients and skin creams are especially a fire safety concern when used by people who spend extended periods in a bed or armchair due to illness or impaired mobility.

Anyone using emollients or skin creams regularly should keep well away from fire, naked flames or heat sources. A build-up of residue on bedding, clothing and dressings can increase flammability.

- Do not smoke.
- Loose clothing can easily catch fire – take care not to lean over a hot hob and roll up your sleeves, if possible, when cooking.

- Keep candles away from clothing, including when lighting them.
- Sit at least 1 m away from a heater – sitting too close could easily set light to clothes or chair. Always follow the manufacturer's instructions when using creams and emollients.

Case study: Care home (2022)

A care home business run by a private health and care provider was ordered to pay a record £937 500 fine after a resident died in a fire while smoking at one of its care homes.

*** Care Services (ANS) Ltd was fined £937 500 for fire safety failings and ordered to pay £104 000 prosecution costs at Southwark Crown Court in 2022. It is the highest ever fine for fire safety breaches in the UK.

London Fire Brigade brought the prosecution against *** under the Regulatory Reform (Fire Safety) Order after it was called to the blaze at the *** care home in Brockley in March 2016.

A 69-year-old wheelchair user at the home died in a fire while smoking unsupervised in a shelter in the garden of the home. A care assistant saw the fire from a first-floor window and called 999 before staff attempted to put the fire out, but sadly the resident died from his injuries.

A subsequent investigation found that although a smoking risk assessment had been carried out, it did not assess his use of emollient creams, which can be flammable if allowed to build up on skin, clothing or bedding.

Additionally, apparent burn marks indicative of previous incidents were found on Mr ***'s clothing after his death, something which care home staff said they had not been aware of. They said that if they had been, they would have ensured more regular checks were made when he was smoking.

The brigade brought the case as there had been a failure on ***'s part to comply with fire safety duties, which placed Mr *** and other residents at risk of death or serious injury in case of fire.

The company pleaded guilty to contravening Article 11 (1), which relates to the management of fire safety measures. Specifically, it accepted that it had failed to

- ensure staff understood the risks from the use of emollient creams
- warn residents using paraffin-based products not to smoke or require precautions to be taken such as the use of a smock or apron
- instruct staff not to leave a resident using paraffin-based products smoking unsupervised
- carry out an individual smoking risk assessment of the resident as normal with the control measures in place.

The Assistant Commissioner for Fire Safety said

> This case is an absolutely tragic example of what the devastating consequences of failure to comply with fire safety regulations can be.

There are several measures which the home could have put in place to mitigate the fire risk which Mr ***'s limited mobility, emollient cream use and smoking posed – but none of these were implemented.

Mr ***'s family should rightly have been able to trust that he would be safe in a care home, when sadly the opposite was true.

Such a large fine highlights the seriousness of (the) company's failure to protect a vulnerable resident in its care.

If there can be anything constructive to come from this, we hope that it will be that anyone who has a legal responsibility for fire safety in a building – whether as a landlord, property manager, care home provider or any other setting – takes note and makes sure they are complying with the law.

What needs to be considered in respect of evacuation procedures for residential and care homes?

Residents in nursing and care homes may need additional assistance to evacuate the building in an emergency. Many may have mobility issues, many may have dementia and be unable to comprehend instructions, many may be confined to bed and many may have sight and hearing impediments. All such disabilities will need to be addressed by the responsible person when completing the fire risk assessment.

Full evacuation to an external place of safety may not be practicable for many residents and an alternative evacuation strategy must be considered and implemented.

In care premises, evacuations fall into three categories.

Single stage

If all residents are deemed independent of help, all residents can evacuate immediately with minimal assistance.

Progressive horizontal

Occasions arise when most residents are dependent on staff assistance for a successful evacuation. This may have to be done in stages, first moving residents from fire zones to safe zones and subsequently to outside of the premises.

Delayed

In some cases, it is potentially dangerous to immediately evacuate certain residents who have complex needs. Instead, they may be allowed to remain in their rooms while the fire is put out. In such circumstances, additional structural precautions may be necessary – for example, fire-resisting walls, doors and ceilings.

Competent advice must be sought when determining which evacuation strategy to implement and the local fire and rescue service may be able to provide assistance.

Escape time and travel distance must also be considered as part of the evacuation strategy.

The time required to evacuate residents will be influenced by a number of factors, including

- the degree of mobility of each resident to be moved
- the level of awareness of each resident, and the level of co-operation that may be expected
- the distance to be travelled to the adjoining protected area
- the number of staff members available to move residents
- the level of training given to staff in moving residents quickly in an emergency
- the need to disconnect any medical devices and to prepare residents for further evacuation.

Checklist

- ensure that fire compartmentation is good and that fire will not spread. This may involve a more invasive fire risk assessment, especially in older buildings (looking above ceiling voids and under floors etc.)
- ensure that you have a comprehensive fire alarm system that conforms to BS 5839 part 1
- ensure there are sufficient trained staff/carers who know how to react in case of fire and who can safely evacuate residents
- ensure staff are trained in using emergency evacuation equipment (slides, evac chairs etc.)
- if using beds to evacuate, ensure that they will fit through openings
- ensure fire safety measures are always in place and maintained (never wedge fire doors open)
- ensure exit routes are always clear and available for use
- fit a sprinkler system
- make sure you know what to do in the event of a power outage that affects your fire detection and alarm system.

What extra steps in relation to fire safety do I need to take in relation to schools, colleges and universities?

A study from Zurich Municipal, a leading insurer of schools in the UK, revealed 480 primary and secondary schools endured fires in 2019, amounting to 40 incidents every month. As a result, almost 20 000 school children had their education impacted or were displaced from their usual school building over the same period.

Further data analysis by Zurich showed that during 2019 over 15 000 m^2 of classroom space was damaged during blazes across 271 primary and 209 secondary schools.

Malfunctioning appliances or equipment, faulty electrics, arson and kitchen blazes are among the leading causes of school fires. Larger fires in schools cost on average £2.8 million to repair and in some cases over £20 million.

The basic principles of managing fire safety apply to educational establishments.

In addition, consider the following.

General

- possibility of arson
- overall number of students
- language ability of students
- mobility of students and visitors
- areas of lone working/studying
- out-of-hours emergency plan
- fire safety training of staff and students.

Classrooms

- keep combustible material away from heaters
- avoid portable heaters
- clearly mark the evacuation route on the school floor plan and place it adjacent to the classroom door
- check electrical appliances and cords regularly and disconnect appliances where practicable
- keep all doorways clear
- dispose of rubbish daily
- protect all unused outlets in kindergarten and pre-school nurseries with safety plugs
- do not use candles or similar open flames in classrooms.

Laboratories

- store flammable liquids properly and dispense them from an approved safety container
- avoid storing incompatible chemicals next to each other as interaction may occur and cause a fire or explosion
- survey chemicals annually and discard any that are obsolete or show signs of decomposition
- order and store the minimum supply necessary
- clean up spills immediately
- repair all electrical hazards and avoid overloading outlets
- dispose of rubbish regularly and in a proper manner.

Gymnasium/auditorium

- check emergency lighting monthly
- keep all doorways clear
- keep exit lights in working order
- extension cords should not be used as a replacement for permanent wiring.

Kitchen and home economics (consumer and life sciences)

- keep all cooking equipment, hoods, filters and ducts free of grease accumulation
- clean up spills immediately
- do not allow garbage to accumulate
- regularly clean lint traps on dryers and filters on vents, fans and air conditioners
- never leave hand irons plugged in. Any electrical outlets used for plugging in irons should be equipped with a pilot light.

Arts and crafts/industrial technology

- electrical outlets and frayed cords can cause burns and shock. Avoid overloading outlets and use approved extension cords only on a temporary basis
- secure compressed gas cylinders in an upright position. Avoid dropping, rolling or knocking them together
- never weld in areas where combustible material is stored
- keep work area clean and check daily for fire hazards
- provide suitable waste cans for disposal of oil-soaked and paint rags
- use safety cans for storing and dispensing small amounts of flammable liquids
- ensure good ventilation for painting operations and a regular routine for cleaning up overspray residues
- supply suitable and easily accessible fire extinguishers.

Office and storage areas

- avoid overloading electrical outlets and keep wiring away from doorways, windows or under carpeting
- store and handle cleaning compounds and polishes carefully (they often give off flammable vapours), providing ventilation where possible
- ensure general order and cleanliness in storage rooms.

Halls and doorways

- identify exits with lighted exit signs and keep them unobstructed
- avoid wedging open hallway doors
- check emergency lighting every month to ensure it is in proper working order

- test fire doors regularly to make certain that the doors open without difficulty and that the hardware is not broken
- restrict paper to bulletin boards, fasten at all corners and do not exceed 20% of a wall area.

Boilers and maintenance rooms

- do not store combustibles in boiler rooms or furnace rooms
- service heating equipment annually and check units regularly to make sure they are in proper working condition
- do not store combustible materials in electrical rooms
- gas-powered equipment should be stored outdoors or in a fire-rated room (1 h fire separation).

Outside areas, cycle sheds, plant rooms and refuse stores

- keep plant and refuse rooms locked
- remove combustible materials from all areas
- keep combustible materials – for example, recycling products – at least 8 m away from any buildings
- consider installing smoke detectors/alarms
- remove all ignition sources
- consider installing CCTV
- consider installing passive infrared lighting systems
- provide easily accessible fire extinguishers.

Case study: School arson attack (2022)

Despite the efforts of 50 firefighters, a large part of Ash Green Primary School in Halifax, West Yorkshire, was destroyed following a large blaze in February 2022.

Much of the primary school was reduced to ash after a huge fire burned it to the ground.

Fifty firefighters battled to control the inferno, but they were unable to save the school's roof.

A 19-year-old man found at the scene by officers was given, in October 2022, a life sentence, to serve a minimum of nine years, for burglary and two arson offences relating to the fire at the school.

What are some of the specific fire safety measures I need to be aware of for entertainment premises?

Fire safety training

Entertainment premises, places of assembly, theatres, cinemas and concert halls all pose complex fire safety issues because, in most premises, large numbers of people gather who are unfamiliar with the premises and any emergency procedures. They have not been trained in emergency evacuation procedures and often will not know what to do. Panic often ensues and people are injured in mass evacuations.

The major priority for entertainment venue operators is the training of staff and others associated with the business in emergency procedures and building evacuation.

It is essential that venue owners/operators carry out regular fire safety training for all staff, implementing where practicable, physical drills to evacuate the building.

Crowd control

Many venues disregard the maximum occupancy number for the building, believing that it is a number that is unimportant and which can be ignored.

The maximum occupancy number is based on a calculation of safe occupancy in relation to the number of fire exit routes and how the building is to be used – for example, standing room only, seating for restaurant services, fixed seating, moveable seating.

Operators must be able to show that they have procedures in place for calculation occupancy levels – for example, door entry number counters – and that they can restrict capacity once the maximum occupancy level has been reached.

Many fatalities and injuries are caused by people surging towards exit routes, causing crushing. People fall and become obstacles to the evacuation of others and so the number of people builds up and evacuation becomes impossible.

People with disabilities

Visitors to venues may have a range of disabilities, some of which could be hidden from general observation.

Responsible persons will need to assess the needs of people with disabilities, especially those with hearing and sight impairments.

Methods of raising the alarm will need to be considered for those with hearing impairment and emergency signage will need to be considered for those with sight impairment.

Safe refuge areas will need to be identified.

Determine the maximum occupancy numbers

Worked example: the Anywhere Tavern, Anytown

The safe occupancy figure for the Anywhere Tavern has been determined at 654 people. The means of escape capacity for the Anywhere Tavern has been determined as 660 people.

Means of escape

The premises operates as a public house with a customer area of approximately 327 m^2.

The pub is laid out to provide both vertical drinking areas (around the bar counter) and customer eating areas, which are provided with tables and chairs.

The premises has an external drinking courtyard.

The safe capacity number for a public entertainment premises must always be equal to or less than the provision made for means of escape – that is, there must always be adequate and safe means of escape from a premises.

The Anywhere Tavern is provided with the following fire exit doors which provide a means of escape to an external place of safety

- double set of main entrance doors and single main entrance door – that is, one triple set
- exit doors to external courtyard and their subsequent fire exit doors – 2 x single doors
- fire exit doors – double width – to the side of the building into the delivery yard
- double fire exit doors to the side of the building – into delivery yard.

Allowing for a 3 min evacuation time, a double set of fire exit doors will facilitate 220 people to exit the building in an emergency.

A single fire door allows 110 people to exit in 3 min.

(NB: minimum door widths apply – 750 mm for single, 1050 mm for double.)

The means of escape doors at the Anywhere Tavern will enable the following to escape in 3 min

3 × double set of doors	= 3 × 220 = 660
3 × single doors	= 3 × 110 = 330
Total	**990**

In accordance with good practice, it is appropriate, however, to *discount* a main set of fire exit doors as they may not all be available in an emergency – that is, the triple set entrance doors.

The **means of escape capacity** for the Anywhere Tavern will therefore be

990 people – 330 people = **660 people**

The safe occupancy figure can therefore be equal to or less than 660 people.

Safe occupancy

To establish a safe occupancy figure which allows public safety and customer comfort, the Building Regulations 2010 Fire Safety: Approved Document B (as amended 2022) allow the following

- bars and vertical drinking areas: 3 people per m^2
- restaurants, dining areas, seating areas: 1 person per m^2

Furniture can be removed to provide more standing area in any public house and so a reasonable assumption would be to assume that 50% of the customer floor space is for vertical or bar drinking and 50% for customer seating for eating and so on.

If such a ratio were considered for the Anywhere Tavern the capacity would be

$$50\% \text{ of } 327\ m^2 = 163\ m^2 \times 3 \text{ people} = 489$$

$$+$$

$$50\% \text{ of } 327\ m^2 = 163\ m^2 \times 1 \text{ person} = 163$$

$$\mathbf{652}$$

The total capacity would be 652 people.

This figure would be acceptable because the means of escape calculated previously caters for 660 people.

Many fire authorities will consider a pragmatic approach to capacity numbers in pubs, and will recommend that the ratio of two people per m^2 is used for calculation purposes.

Applying the ratio of two customers per m^2 of customer area will enable the Anywhere Tavern to accommodate

$$(327\ m^2 \times 2) = 654$$

This is within the means of escape capacity and can be taken as an acceptable capacity.

Conclusion

The final figure for means of escape provision and safe occupancy levels are in accord and so it can be deemed a reasonable, suitable and sufficient assessment that the safe occupancy figure will be **654 people**.

Such figures must include all persons in the premises, including staff.

Case study: Cameron House Hotel fatal fire (2017)

Two men died in a fire at a hotel in Loch Lomond, Scotland, in 2017 after fire spread rapidly through the hotel. The fire began in the concierge cupboard of the hotel after a night porter stored hot embers from an open fireplace in the hotel in the cupboard alongside combustibles.

The owners of Cameron House and the porter were both convicted of safety offences related to the fire in 2021.

Sheriff Thomas McCartney, who led the fatal accident inquiry, determined that there were

> defects in the system of working at the hotel which contributed to the accident resulting in the deaths.

The defects found included

- the careless disposal of ash

- the lack of a written standard operating procedure reinforced by staff training
- the absence of appropriate equipment for safe disposal of ash
- ash bins in the service area being full and the lack of a 'coherent system' to regularly empty them
- the absence of instructions to staff relating what to do with ash when bins were full
- the presence of combustibles within the concierge cupboard, notwithstanding that the fire service had warned that they should be removed.

Other matters were considered relevant to the circumstances of the deaths too.

The guest list was left inside the hotel by the night manager when the hotel was evacuated, requiring a fireman to retrieve it. This delayed efforts to ascertain who was missing.

An independent assessment of fire safety at the hotel was also altered to state that all of the recommendations identified in the previous report had been recorded as complete, without evidence to support that or independent verification.

Voids in the walls and ceiling of the hotel's main building also enabled undetected fire to spread from the cupboard into other parts of the building, the sheriff found.

The building and fire safety standards as they apply to hotels and, in particular, older buildings, were also cited as relevant to the circumstances of the deaths.

An extract from the Sherriff's Fatal Accident Inquiry report noted

> The fire within the hotel continued to develop. Firefighters described seeing flames inside the walls on the first floor above the reception and flames emanating from the third floor. A wooden staircase situated directly above the seat of the fire was still intact. This indicated that the fire had not broken through in an expected manner but had spread via voids above the concierge cupboard. The smoke spread from a fire in a building which does not have voids would be expected to be from one compartment to the adjacent compartment, meaning from the ground floor (where it originated) to the first floor, then up to the second floor.

Fire safety legislation for commercial premises is broadly the same throughout the UK – the Fire (Scotland) Act 2005 and the Fire Safety (Scotland) Regulations 2006 in Scotland and the Regulatory Reform (Fire Safety) Order 2005 in England and Wales. The legislation places duties in respect of fire safety on employers and on those in control of 'relevant premises'. Duties include an obligation to carry out a risk assessment to identify fire safety risks and put in place reasonable mitigation measure to address identified risks. Emergency response arrangements are required.

KEY RESOURCES

Building Regulations 2010 Fire Safety: Approved Document B (as amended 2022)
Fire safety risk assessment: small and medium places of assembly (Gov.uk 2006)
Fire safety risk assessment: large places of assembly (Gov.uk 2006)
Fire safety risk assessment: educational premises (Gov.uk 2006)
Fire safety risk assessment: residential care premises (Gov.uk 2006)

Perry P
ISBN 978-0-7277-6647-2
https://doi.org/10.1680/fsbqa.66472.203

Chapter 13
Fire safety training

What are an employer's responsibilities in respect of training employees in fire safety?

Employers have duties for providing employees with information, instruction and training in respect of all health and safety matters, including fire safety.

Section 2 of the Health and Safety at Work etc. Act 1974 imposes a general duty on employers in respect of providing information, instruction and training.

Regulation 13 of the Management of Health and Safety at Work Regulations 1999 requires employers to provide training for employees as follows

- upon recruitment and induction
- whenever an employee is exposed to new or increased risks from production, processes, plant, premises or people
- whenever training needs to be repeated or to be continuous so that the employee is up to date with current best practice.

Training must be flexible and adaptable to new working practices and consequential risks.

Training must be undertaken during working hours.

Article 21 of the Regulatory Reform (Fire Safety) Order 2005 states

> The responsible person must ensure that his employees are provided with adequate safety training
>
> (*a*) at the time when they are first employed; and
> (*b*) on their being exposed to new or increased risks because of—
>
> (*i*) their being transferred or given a change of responsibilities within the responsible person's undertaking;
> (*ii*) the introduction of new work equipment into, or a change respecting work equipment already in use within, the responsible person's undertaking;
> (*iii*) the introduction of new technology into the responsible person's undertaking; or
> (*iv*) the introduction of a new system of work into, or a change respecting a system of work already in use within, the responsible person's undertaking.

The training referred to in paragraph (1) must

(*a*) include suitable and sufficient instruction and training on the appropriate precautions and actions to be taken by the employee in order to safeguard himself and other relevant persons on the premises;
(*b*) be repeated periodically where appropriate;
(*c*) be adapted to take account of any new or changed risks to the safety of the employees concerned;
(*d*) be provided in a manner appropriate to the risk identified by the risk assessment; and
(*e*) take place during working hours.

Is it necessary to keep training records?

While the need for training records is not specifically laid down in law, it is good practice to keep records of who has had what training and when.

Employers will need to be able to demonstrate that they have complied with their legal duties and the best way is to keep records.

If premises have a premises licence under the Licencing Act 2003 there may be a requirement in the operating schedule attached to the licence for fire safety training to be undertaken and records kept. This may be particularly relevant where employees are expected to assist in the evacuation of members of the public from places of entertainment – for example, cinemas, theatres, nightclubs and events venues.

How often should fire safety training be carried out?

It is recommended that fire safety training is carried out at least twice a year.

The more that staff are trained to be familiar with what to do in an emergency, the greater the chance is that people will be evacuated safely and the threat to life safety reduced.

Fire safety training should always be carried out within the first day of a new employee starting work or when an existing employee is transferred to another part of the building.

Who can give training in fire safety?

The law expects 'competent' persons to be involved in giving advice and carrying out training in health and safety, including fire safety.

There are no specific qualifications needed in order to establish competency in delivering fire safety training, but certain criteria should be demonstrated

- experience in the workplace and the fire safety precautions in place
- knowledge of the fundamentals of fire and how and where it can start, its effects and so on.

It is not always necessary to give practical demonstrations in how to use fire extinguishers as practical videos can be used.

Is there a difference between induction training for fire safety and general fire safety training?

Induction training in fire safety is specified under the Management of Health and Safety at Work Regulations 1999 and generally is intended to cover key emergency information an employee needs to know to ensure their safety.

Induction fire safety should at the very minimum include

- method of raising the alarm
- evacuation procedures
- location of means of escape
- assembly point
- checking in/clocking in procedure
- major fire hazards within the building.

Employees should be given induction training on day one of their employment. It is good practice to develop everyone's basic knowledge by scheduling employees to attend more detailed fire safety training.

I intend to give my employees further fire safety training. What subjects should I cover?

Fire safety training need not be complicated. The following subjects should be included

- the action to take on discovering a fire
- how to raise the alarm and what happens then
- the action to take on hearing the fire alarm
- the procedures for alerting customers and visitors, including where necessary, directing them to exits
- the arrangements for calling the fire brigade
- evacuation procedures for everyone to reach the assembly point at a safe place
- the assembly point
- the location and, where appropriate, the use of firefighting equipment
- the location of the escape routes, especially those not in regular use
- how to open all escape doors, including the use of any emergency fastenings – for example, break glass bolts
- the importance of keeping fire doors closed to prevent the spread of fire, heat and smoke
- how to isolate machines, equipment and so on, as necessary
- how to isolate power supply sources
- why not to use lifts (if provided)
- evacuation of disabled customers or staff
- importance of general fire safety and good housekeeping
- understanding the fire risk assessment, where it is kept, when it needs to be updated and so on

- the control measures in place in the premises
- the need to report equipment faults and malfunctions
- how to deal with spillages.

Will fire wardens need additional fire safety training to that which they have had as an employee?

The topics for the fire warden or marshal fire safety course will probably be similar to those of any fire safety course but the level of information that they may need will be greater.

Fire wardens should be very aware of what their role is and must be clear about their responsibilities. So, the first aspect of fire warden training will be to state clearly that their role should be

- to take appropriate and effective action if a fire occurs
- to ensure that escape routes are kept available for immediate use
- to identify and report fire hazards in the workplace.

The syllabus should then include

- fire safety management within the company
    - the company fire safety policy
    - the fire warden's role
    - developing a 'fire safe' environment
    - company fire procedures
- the nature of fire
    - the triangle of fire
- means of escape
    - fire warning
    - escape routes
    - emergency exits
    - signage
    - fire evacuation
    - fire drills
- fire prevention
    - good housekeeping
    - arson spotting

- firefighting
  - location and type of fire extinguishers
  - how to use them
  - which fire extinguisher is used on what type of fires
  - fire blankets
  - use of sprinklers
  - how to identify fire extinguishers
- fire safety monitoring
  - fire safety checks
  - hazard spotting
  - fire prevention – electrical testing of appliances and so on.

What is the requirement for training in fire drills?

A practice fire drill should be carried out at least once every twelve months in most premises, and in those which are high risk or open to large numbers of the public – for example, department stores – fire drills should be carried out more frequently.

Fire escape routes should all be available for practice fire drills except that it is good practice to assume that one is inaccessible, as would be likely in the event of a fire.

Even if members of the public are not present for the fire drill, it is sensible to carry one out because it is important that the staff know what to do in an emergency and that they are trained and familiar with the steps to be taken.

The full fire alarm should be activated by a member of staff as if it were an emergency and the procedures followed should be those that would be practised in a real emergency.

The effectiveness of a fire drill should be observed and monitored for efficiency and effectiveness. It is of little value if a fire drill is carried out that is ineffective and remedial measures are not taken to put it right.

It is important to check the time it takes to evacuate different parts of the building – for example, floors – and the time it takes to evacuate the whole building.

- Is this in accordance with any times stipulated in the fire risk assessment? Is it reasonable? Does it leave people vulnerable?
- Did people go to a place of relative safety? What happened to them once there? Were they assisted in the final exit?

- Did everyone go to the assembly point? Was anyone missing who should have been accounted for?
- Did fire wardens check their floors/areas?
- What happened to anyone with any disabilities? Could they have been effectively evacuated?

Do employers have duties to train people other than their employees?

Employers must provide appropriate information, instruction and training to those who are affected by their business or undertaking.

Obviously, members of the public cannot be trained in fire safety, but they can be provided with information such as fire safety signs and so on.

Visitors to any business premises should be informed or instructed in fire safety procedures. This is often done at the reception and visitors are given name badges and brief information on what to do in the event of a fire. This can be on the reverse of the ID card or a separate instruction displayed on the counter.

Contractors working on the premises should be treated differently and they must be given information and instruction about the fire hazards and risks associated with the workplace. They must also be given information on emergency procedures, the fire alarm system and so on.

In addition, contractors must give information to the employer about any tasks that they are going to undertake that may cause hazards and risks to the employer's workforce.

Any use of hot works must be discussed and, where necessary, it would be good practice to run through a 'toolbox' talk on the fire hazards, risks and control procedures for using hot works, including any hot work permit procedure.

The overlap between employees working on the premises and contractors working on the premises in relation to health and safety is merging and it is now common practice for contractors and their employees/operatives to be invited to attend employer health and safety training courses.

As most fires in commercial buildings are caused during refurbishment works, it makes good business sense to include contractors in fire safety training.

Should employees be trained in managing fire safety for people with disabilities?

Yes. A responsible person, often the employer, has a legal duty to ensure the safety of all persons in the event of fire.

Special provision may need to be made to ensure that people with disabilities are not prejudiced in the event of a fire and they may need assistance when fire alarms sound or when an evacuation is necessary. Employees and others will need to know how to respond in the event of an emergency and will need to know what assistance to give a person with disabilities without jeopardising their own or others' safety.

Every person with a disability should have a personal emergency evacuation plan (PEEP) that has been designed specifically for them and their unique disability. The individual must be familiar with the content of their plan and must be practiced in following it, and training is essential.

Persons who are responsible for assisting people with disabilities, often called buddies, must also be familiar with each individual PEEP, and training is also essential.

Where personal aids are used – for example, evacuation chairs, vibrating alarms, audible alarms and so on – employees or others responsible for managing the evacuation of persons with disabilities must have suitable training in the use of such equipment.

Are there any competency standards relating to the training of fire safety?

There are no specific competency standards relating to fire safety training but, as with all health and safety training, it should be delivered by someone who is competent to do so – that is, has the skills, knowledge and experience to convey the key messages about the risks from fire and the procedures to be followed to reduce those risks.

The responsible person should ensure that all fire safety training complies with legal requirements and that it covers the topics necessary to meet the standard that 'every employee and other persons has received information, instruction and training as is necessary to carry out their duties safely'.

Responsible persons should set out their own competency standards for fire safety trainers – for example

- they must have industry/sector specific knowledge of fire safety hazards and risks
- they must show evidence of professional fire safety qualifications
- they must evidence training experience and preferably training qualifications for the trainers appointed
- they must provide suitable references
- they must understand the organisation they are training and be familiar with site specific procedures.

KEY RESOURCES

Regulatory Reform (Fire Safety) Order 2005
Fire Safety Act 2021
Fire safety law and guidance documents for business – www.gov.uk – search for fire safety
Management of Health and Safety at Work Regulations 1999

TOP TIPS

- keep training simple so that key messages can be understood
- try to show a short fire safety video as this can aid learning
- keep records, either electronically or in paper form
- train all new staff as soon as possible
- undertake refresher training on a regular basis
- train staff if anything changes in the premises – for example, new extensions, internal alterations, increased storage of materials or use of flammable materials and so on
- make sure staff know about the fire risk assessments
- train employees on your emergency plan
- instruct or train contractors working on the premises
- train employees in assisting people with disabilities in the event of a fire.

CHECKLIST FOR FIRE WARDENS

If a fire is discovered, the fire warden should

- ensure the fire alarm has been activated
- check that equipment and processes are shut down, where safe to do so
- evacuate people from the area they are responsible for or from the whole building, but without putting themselves at risk
- advise everyone to go to the assembly point
- assist any employees or visitors with disabilities and ensure others help with their evacuation
- report to the 'chief' fire warden that all people are accounted for or if anyone is missing
- report if any areas of their floor or building were not checked for people – for example, any remote working areas
- guide employees back into the building when advised by the fire brigade it is safe to do so.

Fire wardens are *not* expected to fight the fire unless it is manageable and they feel competent to do so as a result of training.

FIRE SAFETY
STAFF INSTRUCTION AND TRAINING RECORD
GENERAL RECORD

Name of employee	Date of training	Who gave the training?	What information was covered?	Signed (employee)

EMPLOYEE RECORD FOR FIRE SAFETY TRAINING

Name of employee: ______________________________
Job title: ______________________________
Date of training: ______________________________
Duration of training: ______________________________
Training given by: ______________________________

Subjects covered	**Yes**	**No**
Fire hazards		
Safe practices		
How to raise fire alarm		
Identifying the fire alarm sound		
Action to be taken on hearing alarm		
Calling the fire brigade		
Names and duties of fire wardens		
Location of firefighting equipment		
How to use equipment		
Types of other fire protection		
Escape routes to be used		
Fire doors – need to be kept shut		
Location of assembly point		
What *not* to do in a fire		
Assisting people with disabilities		
Daily visual fire checks		
Reporting hazards, dangers etc.		
Other: please specify		

Signed by employee: ______________________________
Signed by instructor: ______________________________

FIRE DRILL RECORD

Should be undertaken on a quarterly basis

Date	Was all the building evacuated? Yes/No	Did the drill include customers? Yes/No	How long did it take to empty the building (minutes)?	What do you need to do to improve anything for next time?

Perry P
ISBN 978-0-7277-6647-2
https://doi.org/10.1680/fsbqa.66472.215

Chapter 14
Fire safety for people with disabilities

As an employer, am I legally responsible for the safety of people with disabilities?

Yes. The law requires employers to be responsible for the health, safety and welfare of all employees and to make whatever special arrangements are necessary to deal with any employees who have 'special needs'.

Also, an employer has duties to persons not in his employ and must ensure that their safety is not put at risk by his undertaking.

Persons in control of premises must also consider the safety of persons 'resorting to the premises'.

Is there any other legislation applicable to people with disabilities?

The Equality Act 2010 applies to all employers and service providers and requires them to treat people with disabilities 'no less favourably' than able-bodied persons.

Employers must not discriminate against people with disabilities and must ensure that adequate facilities are provided for them while they are at work. This will include procedures for fire safety, fire evacuation and so on.

Service providers include everyone who offers a service of some kind to the public or others and includes shops, banks, professional organisations, lawyers, insurance brokers, libraries, entertainment premises, offices, recreational buildings and so on.

Service providers must provide their service to no less a standard to people with disabilities than it does to non-disabled persons.

People with disabilities must have safe and reasonable access to goods and services. If there are physical barriers to access, then the service provider must make 'reasonable adjustments' to accommodate them.

What are disabilities?

Disabilities include physical impairment, lack of mobility, impaired hearing, impaired vision, mental illnesses and any condition that can be judged to have a long-term debilitating effect on the individual. The code of practice supporting the Equality Act 2010 lists examples of disabilities. The list is not exhaustive and if any person believes that they have a disability and have been discriminated against they can apply to the Equality Commission for a judgement.

How do disabilities – both physical and other – affect fire safety requirements?

An employer and service provider must not discriminate against people with disabilities and this will also include in any emergency procedures.

Fire evacuation procedures, means of escape and so on must address the needs of people with disabilities.

The fire risk assessment required by the Regulatory Reform (Fire Safety) Order 2005 must consider any additional fire safety risk to people with disabilities.

Control measures must be adopted that do not discriminate.

What special consideration needs to be given to people with disabilities in respect of fire safety?

The special considerations needed for people with disabilities are contained in British Standards BS 9999:2017 and BS 9991:2015. (NB: a new edition BS 9991:2021 is currently under review by BSI and is due to be published in late 2023.)

In summary, the following need to be considered in the fire risk assessment.

Means of escape

Door widths should be a minimum of 900 mm. Corridor widths need to be wide enough for wheelchair use with helper – approximately 1000 mm. Escape routes must be kept clear and properly signed. Any exit route especially designed for wheelchairs should be signed with a wheelchair symbol.

Escape routes should be on the same level with no steps. Ramps should be minimal and should not provide gradients steeper than 1 in 12.

Handrails should be provided along escape routes where practicable.

Disabled refuges

A refuge is a place of relative safety and is provided where people with disabilities may need to wait for further assistance – for example, because they are unable to descend the stairs to the final exit.

A refuge must be in a fire protected area and separated from any fire by a fire-resisting construction.

A refuge may be

- a protected lobby
- half landing on a staircase
- an open-air area such as a flat roof, podium or similar that is sufficiently protected or remote from any fire risk, and is provided with its own means of escape and protected from the risk of falls.

It is advisable to consult with the fire authority on the location and use of disabled refuges.

People waiting in a disabled refuge must not cause obstructions to others exiting the premises.

A method of recording who is likely to be in the refuge must be in place so that the fire brigade can be advised immediately on their arrival to the emergency.

Fire alarms

People with impaired hearing may not be able to hear fire alarms and consideration must be given to providing flashing lights or visual fire alarms.

Again, the fire risk assessment should consider the different types of alarms. Sometimes the high pitch fire alarms are audible to some people with impaired hearing.

Fire signage

People with impaired sight may not be able to find their way out as they cannot see signs.

All signs must be in large lettering and easily visible from all parts of the building they serve.

Good lighting is essential and emergency lighting or photoluminescent signs may assist a person deciphering instructions.

Colour contrasts may help people with impaired vision.

Handrails, steps and so on

Handrails should be 'tactile' and protrude beyond the end of the fire escape route or stairs (without causing obstruction).

Steps and nosings should be colour-contrasted – for example, white nosings, as these make edges clear.

Tactile flooring may be suitable in front of exit routes or on ramps and so on.

Fire evacuation notices

Instructions for how to respond in the event of a fire must be available for people with disabilities. It may be possible to give people with impaired vision Braille fire alarm notices, or those with impaired hearing vibrating alarms.

Audio tapes may provide verbal instructions.

The use of pictogram signs helps people easily to see which emergency route to take and so on.

The emergency plan

The emergency plan should include details of what procedures are to be followed in the event of an emergency, and specific details for actions to be taken for people with disabilities must be included.

The plan should include details of what staff should do when assisting in the evacuation of people with disabilities – for example, a sighted person leading the visually impaired.

Consideration should be given to how guide dogs for the blind and hard of hearing will be dealt with.

Lifts should not be used unless designed specifically for use in a fire – that is, firefighting lifts.

Premises providing residential accommodation must have detailed emergency plans for evacuating people with disabilities.

Case study: Hotel

A hotel company was required by the Equality Commission to compensate a guest who felt discriminated against during a fire evacuation as he did not receive any specific information or assistance in evacuating his hotel room and was left unattended in what could have been a burning hotel. The hotel had to revise its procedures.

What are personal evacuation plans (PEEP)?

The aim of a PEEP is to provide people, who cannot get themselves out of a building unaided, with the necessary information and assistance to be able to manage their escape to a place of safety, and to ensure that the correct level of assistance is always available.

It is the responsibility of a manager, responsible person or nominated representative to talk to staff, service users and visitors with disabilities to assess whether they require any assistance in the event of an emergency. If someone believes they might require assistance, a PEEP questionnaire should be completed. This should be completed by the responsible person or their nominated representative and, if possible, in liaison with the individual who has the disability or mobility impairment.

There are many examples of PEEP questionnaires or templates, and a couple of examples are included at the end of this chapter.

Are PEEPs required for residents in high-rise residential buildings?

The government have been considering the need for PEEPs for residents of high-rise buildings after the Grenfell Tower Inquiry report recommended that such plans should be put in place in all higher-risk buildings.

However, it is unlikely that legal requirements will be put in place as responses to a consultation paper indicated that implementing such a legal requirement would be difficult due to practicality and proportionality.

The government decided to consult on additional measures during 2022 involving emergency evacuation information sharing and expect to issue a response indicating their preferred options during 2023.

Many social housing providers have taken the decision to develop PEEPs for their tenants in higher-risk buildings so that there is a clear plan of action in the event of any evacuation of the building.

Is a 'stay put' policy suitable for people with disabilities?

A 'stay put' policy is an evacuation strategy used in purpose-built blocks of flats. When a 'stay put' policy is implemented, it is to keep people safe when they are not in an area directly affected by the fire.

Stay put policies may be suitable for people with disabilities but they must be considered in the wider context of the individual PEEP.

When a fire is not in their flat, they should stay inside with the doors and the windows closed. Usually, it is considered a suitable safe environment for up to one hour due to the construction of the building and the compartmentation.

If a fire should occur in a person's flat or common areas of a building, they would be advised to evacuate and call the fire and rescue services.

Stay put policies should not be adopted as an easy way to deal with the evacuation of people with disabilities.

Responsible persons should not rely on the fire and rescue service to evacuate vulnerable people from their building.

PEEP QUESTIONNAIRE

Personal information

Name

____

Location

1. Which part of the building do you normally work in / visit?

Please name: the building, the floor and the room.

____

____

2. Do you routinely use more than one location in this building? YES / NO

If you feel it is necessary, please provide further details below. (Please list the floors and rooms you use in each building.)

____

____

____

____

____

____

____

____

____

____

Emergency procedures

Are you aware of the emergency evacuation procedures that operate in the building(s) you attend? YES / NO

Do you require written emergency evacuation procedures? YES / NO

Do you require written emergency procedures to be supported by British Sign Language interpretation? YES / NO

Do you require the emergency evacuation procedures in Braille? YES / NO

Do you require the emergency evacuation procedure to be in digital or audio format? YES / NO

Do you require the emergency evacuation procedures to be in large print? YES / NO

Are the signs that mark emergency routes and exits clear enough? YES / NO / Don't know

Emergency alarm

Can you hear the fire alarm(s) in your place(s) of work? YES / NO / Don't know

Could you raise the alarm if you discovered a fire? YES / NO / Don't know

Assistance

Do you need assistance to get out of your place of work in an emergency?
YES / NO

Is anyone designated to assist you in an emergency? YES / NO / Don't know

If YES give name(s) and location(s)

____

____

Is the arrangement with your assistant(s) a formal arrangement? (A formal arrangement is one specified for them by a senior person or written into their job description or by some other procedure.) YES / NO / Don't know

Are you always in easy contact with those designated to help you? YES / NO

In an emergency, could you contact the person(s) in charge of evacuating the building(s) in which you work and tell them where you were located? YES / NO

Getting out

Can you move quickly in the event of an emergency? YES / NO

Do you find stairs difficult to use? YES / NO

Are you a wheelchair user? YES / NO

The information you have given us will help us to meet any needs for information or assistance you may have.

Please return to:

____

KEY RESOURCES

Regulatory Reform (Fire Safety) Order 2005

BS 9999:2017 Fire safety in the design, management and use of buildings. Code of practice

BS 9991:2015 Fire safety in the design, management and use of residential buildings. Code of practice – under review 2023

Fire Safety: An Employers Guide (Home Office) Equality Act 2010

Guidance: Fire safety risk assessment: means of escape for disabled people. 2007: Gov.uk

Guidance to support a temporary change to a simultaneous evacuation strategy in purpose-built blocks of flats. 2022: Fire Protection Association

Perry P
ISBN 978-0-7277-6647-2
https://doi.org/10.1680/fsbqa.66472.223

Index

Printed and bound by CPI Group (UK) Ltd, Croydon, CR0 4YY

15/09/2026

14962488-0007